Masterworks

of Technology

"From inventing the wheel to mastering rocket science, E. E. Lewis highlights the joys and challenges of successful engineering. *Masterworks of Technology* is an interesting read for engineers and inventors everywhere—and the people who live with them!"
—Tom Kelley, General Manager, IDEO; Author, *The Art of Innovation*

"This is a delightful, instructive, and well-written book that bridges the two cultures between art and science in a remarkable way. We are taken on a ride through history from Stonehenge to the Apollo program, with visits to the design of the wheel, the construction of a cathedral, the development of the steam engine, the industrial revolution, the Model T Ford, and the Wright flyer. I never knew industrial history could be so interesting. This book could form the basis for a TV program and should certainly be read in schools and universities. It is delightful and the author, who is a distinguished nuclear engineer, should be congratulated."
—M. M. R. Williams, Emeritus Professor, University of London;
Editor, *Annals of Nuclear Energy*

"Lewis leads a fascinating journey through five thousand years of engineering and related sciences. His special insights into the physical and technological challenges facing humanity fire the imagination and create a compelling vision of progress."
—Bob Barnett, Executive Vice President, Motorola

"Elmer Lewis has stitched together many disparate sources of knowledge about human technological progress to create a unique mosaic that eloquently captures the history of innovation and invention. He also gives expression to those critical motivating forces that many academics fail to recognize, let alone grasp: curiosity, necessity, and enlightened self-interest."
—James Conley, Professor, Kellogg School of Management,
Northwestern University

"*Masterworks of Technology* is written to explain to the layperson what engineering, architecture, and design are all about, using historical examples ranging from the pyramids, the wagon wheel, cathedrals, the steam engine, and the telegraph to the modern jet airliner and the computer. The primacy in engineering of design, testing, and the manufacturing process is explained along with the rise of modern science, providing better analytical bases to supplement practical or tacit knowledge for engineering design. Professor Lewis also points out the frequently ignored origins of new scientific knowledge in the consideration of how the products of engineering work. (The original concepts of thermodynamics come from analysis of the steam engine.) Finally, the modern concepts of venture capital and organized innovation are considered.

"I highly recommend this book to young engineering students (and their parents) as well as to the lay public desiring to understand the how the artificial world we all live in is created."
—Donald N. Frey, Former Group Vice President,
Product Development, Ford Motor Co.;
Retired CEO, Bell & Howell Co.;
Professor, Industrial Engineering, Northwestern University

Masterworks

of Technology

The
Story of
Creative
Engineering,
Architecture,
and Design

E. E. Lewis

Prometheus Books

59 John Glenn Drive
Amherst, New York 14228-2197

Published 2004 by Prometheus Books

Inquiries should be addressed to
Prometheus Books
59 John Glenn Drive
Amherst, New York 14228–2197
VOICE: 716–691–0133, ext. 207
FAX: 716–564–2711
WWW.PROMETHEUSBOOKS.COM

08 07 06 05 04   5 4 3 2 1

Library of Congress Cataloging-in-Publication Data

Lewis, E. E. (Elmer Eugene), 1938–
    Masterworks of technology : the story of creative engineering, architecture, and design / E.E. Lewis.
        p. cm.
    Includes bibliographical references and index.
    ISBN 1–59102–243–6 (hardcover : alk. paper)
    1. Technology—History. 2. Technological innovations—History.
3. Technology and civilization. I. Title.

T15.L474 2004
609—dc22

2004011155

Printed in the United States of America on acid-free paper

To Freddy and Robby

Contents

Preface

An unanticipated knock on my door often sets off an experience that is common among my colleagues. Welcomed to my office is a high school student, accompanied by parents, who is considering a career in engineering. Our discussion centers initially around the university's program and admission requirements, the student's academic background and outside interests, and such. Hesitantly, however, an underlying question is raised, frequently by a parent: just what is engineering—what is it that engineers do, and how does it differ from science, if indeed it does?

These are important questions, not only for those contemplating careers in engineering but also for those interested in understanding the nature and impact of technology on society. But answers are not as simple as a dictionary definition. They require understanding the essence of a profession that creates technology ranging from microprocessors to skyscrapers, from consumer products produced by the million to unique, continent-spanning telecommunications systems. The explanations must encompass the profession's many subdisciplines, its interrelations with architecture and industrial design, and, most of all, it must clarify the close and multifaceted interactions between science and engineering.

In fact, one could write a book describing the human endeavor to create technology, and that's what I've attempted to do. The task has required that I step back from my own professional experiences, and from those of my contemporaries, to place the engineering of the twenty-first century in historical perspective and look to the past—to the days of the artisans, to the emergence of modern science—and finally to the future.

In the past, technology was far less advanced and had little to do with the science of its day: civil, mechanical, and other subdisciplines of engineering were still in the future, and no distinction existed between engineer, architect, and designer. And yet the habits of mind and processes of design embedded in today's engineering were already evident. The development of the profession is a fascinating tale, with roots in ancient craft traditions and a heritage enriched by the pioneering efforts of anonymous engineers—monks, shipwrights, cathedral builders, and others—who practiced through the centuries before the partnership between science and technology was forged. Appreciating the transformations brought by the engineering practices of Leonardo da Vinci, Galileo Galilei, and historic figures of later centuries enhances our understanding of what lies at the core of today's practice. For me, such history is indispensable in understanding the process by which engineers combine scientific knowledge, practical know-how, and human values to create the technologies of tomorrow, which we will also explore.

In attempting to capture the essence of engineering, I've had much help from colleagues, friends, and former students. Roger N. Blomquist, J. Edward Colgate, Donald N. Frey, Richard M. Lueptow, and Warren F. Miller Jr. read earlier drafts of the entire manuscript and provided valuable input. Ted Belytschko, Franz Boehm, Michael I. Epstein, James Hand, Giuseppe Palmiotti, Thomas C. Rochow, and Hartmut Wider also pitched in, but any errors, omissions, or other shortcomings found in the pages that follow are my own. The resources of the Northwestern University Library and its helpful staff have been indispensable in bringing this project to fruition; and my years of collegial discussions with Northwestern's faculty and the privilege of teaching its students have been essential ingredients to the book's preparation. My gratitude is also extended in memory of the two scholars whose tutelage most whetted my interest in this endeavor: Prof. Richard Hartenburg of Northwestern and Dr. Willi Beck of Stuttgart. I'm also indebted to my agent, Edward Knappman, most particularly for bringing me in contact with Linda Regan; an author could have no better editor.

More than most, this project has been a long-term family affair. For years, my wife, Ann, has graciously endured the countless leisure hours lost to my preparations for this book, and she and our two children did not object to having family travels detoured in directions that only an engineer would choose. More recently, Elizabeth, now a historian, and Paul, a physicist, have joined Ann in doing their best to prevent my wandering too far astray in attempts to relate engineering to science and society. What follows is dedicated to the next generation, to the two little grandsons who bring so much joy to my wife and me.

Elmer E. Lewis
Evanston, Illinois
2004

One

Creating and Comprehending

R elief came over us as our bodies welcomed the cooler temperatures and our eyes adjusted to the dim light that stood in sharp contrast to the heat and intensity of the summer sun. Relief turned to claustrophobia, however, as we descended from the desert's vast openness farther into the narrow subterranean passage and crawled beneath its low ceiling in procession behind our guide, with my wife in the lead, followed by our daughter and son, as I brought up the rear. I felt as if we were crawling toward the beginnings of history as we entered an inner gallery of Egypt's Step Pyramid.

Reaching the chamber, we stood and stretched our legs as the guide began his recitation of the pyramid's history, construction, and decoration. He explained that the structure that encased us was massive but not nearly as large as the pyramids that adorn so many travel brochures. Nor is the Step Pyramid a true pyramid, for its base is oblong rather than a square, and its 204-foot height is made up of six layers, or steps. But it is Egypt's oldest, dating from the reign of the pharaoh Zoser, in the twenty-eighth century BCE. It marked the beginning of a great age of pyramid building that culminated less than a century later, when the sheer, smooth sides of the perfectly shaped Great Pyramid of Giza reached a height not to be exceeded by another engineered structure for well over four millennia.

As we squirmed back through the passage toward the desert's light, our experience whetted my curiosity to know more about the engineering that went into the pyramids' construction and the ancient engineers who created them. The mastermind behind the Step Pyramid, in fact, is the earliest engineer for whom we have a name. Imhotep was not only pyramid engineer and architect but also adviser and confidant to the Pharaoh. He lived at a time when tempered copper tools had only recently come into being, before the emergence of iron cutting tools, and before wheeled vehicles or even pulleys were known in Egypt. And yet with only tools of wood, copper, and stone, Imhotep's contemporaries brought into being the age of hard rock construction. They abandoned bricks baked from clay and straw and learned to quarry and cut granite blocks from bedrock. They mastered transporting them on the Nile and raising them to the unprecedented heights their creations required. As they gained experience and knowledge, the size, number, and workmanship of the blocks increased. The Great Pyramid at Giza contains more than two million stone blocks, fit together with precision, each weighing in excess of two tons.

Without doubt, Imhotep and his successors possessed great organizational skills in managing labor forces thought to number as many as one hundred thousand. By modern standards, the conditions suffered by the multitudes of conscripted peasants who labored to build the pyramids seem inhumane. But the marvels that resulted from their labor, as well as from the acumen of the ancient engineers, capture the imagination of onlookers even today. The instruments and mathematics available to the Egyptians were exceedingly limited, but they achieved amazing levels of precision in surveying and construction. The accurate leveling of the Giza pyramid's perfectly square base and the precision of its orientation to the points of the compass are extraordinary even by modern standards. But the most striking attribute of the pharaohs' engineers is found in the daring in their design, as they strove to create larger and more perfect pyramids. They faced failure, learned from their mistakes, and went on to create a succession of ever-more-impressive structures.

Following the Step Pyramid the Egyptians went on to build the pyramid at Meidum, a true pyramid with a square base and sides sloping steeply upward at an angle of 55 degrees. It would have reached a height of more than 300 feet had disaster not struck. Technological capabilities were not yet up to the task, and the pyramid collapsed, leaving only a ruin of the central core surrounded by a vast mass of rubble. But the Egyptians persisted. Next came the Bent Pyramid, with a still larger base, and whose sides also would ascend at the 55-degree angle in anticipation of reaching a height of 433 feet. As indicated in figure 1, however, halfway up the angle abruptly decreases to a shallower 43 degrees, creating a flattened apex only 333 feet above the base. Since it seems likely that the Bent Pyramid was under construction at the same time as the collapse at Meidum, the angle correction arguably is the result of the engineers' revision to a less audacious design, which is typical of the added caution that frequently follows catastrophic failure. The 43-degree ascent was retained in the Red Pyramid, the next to be built. But as Egyptian engineers gained knowledge and experience, they successfully increased the slope to nearly 52 degrees and created their largest masterpiece, the Great Pyramid of Giza, which rises to a height of 485 feet. Certainly, this sequence of pyramids exemplifies the lessons learned from failure that Henry Petroski argues are essential to the advance of engineering.[1]

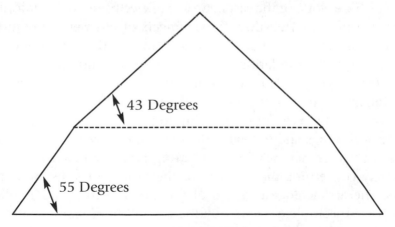

Figure 1. Profile of the Bent Pyramid. (Diagram by Michael Epstein.)

The daring failures, redesign, and subsequent successes of the pharaohs' engineers are among the earliest for which we have evidence, standing today in the form of the pyramids themselves. Other monuments to engineering prowess have survived across the centuries to enrich our heritage: Greek temples, Roman aqueducts, the Great Wall of China, and more remind us of the engineering of earlier ages and of the labors of the slaves, serfs, and peasants who built them. Medieval waterwheels, windmills, early mechanical clocks, great sailing ships, and other technologies have not survived the ravages of time nearly as well. Fire, rot, and corrosion have taken a great toll, but artifacts fashioned from wood and metal are also tributes to the engineering acumen of their time.

The history of technology is fascinating, a never-ending parade of humanity's material products and the uses to which they were put. But more fascinating still than the objects themselves are the motivations and methods that turned ideas to artifacts. Examining evidence of this quintessential human enterprise of designing and building what previously had not existed engenders an appreciation of the age-old technological exuberance for forming new material environments that serve human needs and desires and that produce wealth and power. Superficially, dated methods and centuries-old knowledge would seem to bear little resemblance to the engineering of the aircraft and autos in which we travel, the high-rise buildings in which we work, or the vast networks of electronic communication that tie our world together. The engineers of old were without the power machinery, computer analysis, or many other aids to modern technology. But similarities in habits of mind and approaches to problem solving thread through the history of the engineering profession and remain at the roots of its practice today.

Tracing the techniques of technological innovation from their early beginnings and following their evolution toward today's engineering can sometimes be a frustrating task, for much remains a mystery concerning the practices of the profession's distant ancestors. The archaeological record of humanity's earliest accomplishments is incomplete, skewed by the durability of stone relative to metal, wood, and other materials. Many triumphs have crumbled and disappeared, and the historical record is even spottier. A few books—the volumes of Vitruvius, a Roman engineer and architect, for example—have survived from antiquity. Unfortunately their all-important illustrations have been lost. The scribes who so meticulously reproduced the written word apparently had insufficient aptitude or interest in the more difficult task of reproducing the drawings. Moreover, through the Middle Ages and later, the shipwrights, stonemasons, and members of other guilds carefully guarded their methods, particularly the guild books in which they were recorded. Not until the late medieval notebook of Villard de Honnecourt and the later outpouring of Renaissance drawings can we examine engineering illustrations and appreciate how important visualization has been to engineering. Consequently, we know less of how engineers of earlier times worked than of what they were able to achieve.

Within the fog of technology's earlier history the question arises: what is an engineer? At what historical point can we claim that the artisans, craft workers, or others acceded to the level of practicing what we consider to be engineering? Likewise, at what point in time did the distinctions between architect, engineer, and industrial designer become sharp enough to merit attention? These questions are problematic but perhaps less important than they may at first appear. For today's professions evolved from a variety of roots; the dynamic of that evolution is much more fascinating than determining at what historical point the use of the term *engineer* becomes defensible.

Nevertheless, even today, defining engineering precisely and delineating its professional boundaries remain difficult, for great overlaps exist between it and related endeavors. Much of our attention will be directed toward examining the multifaceted relationships between engineering and science. In addition, the work of architects and industrial designers often is closely entwined with that of engineers. Architecture is centered about one-of-a-kind buildings and other structures, whereas industrial design deals with mass-produced products. But architects' buildings must stand, and industrial designers' products must work, often making the work of architects, industrial designers, and engineers nearly inseparable.

The word *engineer* is derived from Latin *ingeniatorem*, signifying one who is ingenious at devising; whereas *technology* originates from the Greek words *technē*, meaning art or skill, and *logia*, which connotes science or study. In what follows, these definitions are interpreted very broadly, presenting engineers as encompassing those who create—that is, who design and build—technology.

Creators of the technology of earlier ages had only tenuous links to the science—or *natural philosophy*, as it was called—of their times. Certainly long before the advent of the compass, the pyramid builders utilized astronomical or solar observations to square and orient their structures, and they knew how to calculate angles, areas, and volumes. The roots of Western science, so closely associated with classical Greek civilization, were heavily slanted toward geometry and toward the observation of planetary motion and other astronomical phenomena. Throughout antiquity, and indeed until well into the fifteenth century, mathematicians and natural philosophers held the practical arts practiced by the creators of technology in low esteem. With rare exception, they eschewed experiment, since the practical skill that it would require fell far beneath their social status. And the few who did invent clever mechanical devices employed them predominantly for the amusement of aristocratic patrons, with little thought of the useful machinery to which they might lead.

But the lack of a nexus to the sciences of earlier times did not stymie engineering achievement. Roman engineers designed and built buildings, bridges, catapults, and galleons at a time when natural philosophers speculated that an object's shape would determine whether it would float or sink. Medieval engineers harnessed the power of falling water and applied it to numerous labor-saving devices in an age when the dominant Aristotelian philosophy asserted that the speed of falling bodies varies in proportion to their weight. Likewise, technology advanced in China and the Middle East with little benefit from scientific theory. Engineers succeeded in their

quests long before Galileo or Newton. The products of their ingenuity affected everyday life even while natural philosophers' speculations had little effect on the practical arts. Thus nascent science
remained estranged from the technological goals toward which
engineers strove.

China, India, and the Middle East contributed greatly to the
technologies of earlier times. During Europe's middle ages, science
and invention advanced in Islamic lands, and technology in China
reigned supreme. In developing porcelain, metallurgy, canal transport, bridge design, shipbuilding, and more, the Chinese forged
ahead, on sea as well as on land. Early in the fifteenth century the
great junks of the Chinese navy would have dwarfed the ships in
which Columbus later made his voyages. Yet, the subsequent development of engineering from its ancient roots to its present state is a
story concentrated in the West: for while technology languished in
the East, European engineers continued to borrow, adapt, and innovate. The Renaissance, the scientific revolution, and the industrial
revolution brought unprecedented advances in knowledge, tools,
and ingenuity to Western engineers. Visualization, production
organization, and, even more, the emergence of effective methods
for scientific investigation—the melding of controlled experiment
with mathematical analysis—transformed engineering from its
craft-based traditions to the profession that we know today.

The first impact of modern science on engineering wasn't
through the discovery of new phenomena from which resulted revolutionary new technology; that would come later. The impact was
through the systematic investigation of existing technology. Experiments and analysis of man-made artifacts led engineers to replace
blind trial and error with design rules and predictions that allowed
them to improve the strength of their structures and the efficiency of
their machines. From Galileo's analysis of the sagging beams of the
Venetian shipyards through John Smeaton's analysis of waterwheel
and windmill efficiencies well over a century later, the development
of the science of analyzing and improving existing technology had
profound implications for engineering. The development of such
experimentation and analysis, moreover, contributed to science as

well. For often the engineers' attempts to analyze technological problems uncovered poorly understood phenomena, the study of which led to advances in basic science. Ranging from Benjamin Robins's eighteenth-century discovery of the sound barrier from his attempts to predict cannonball trajectories to the understanding of energy and its conservation rooted in nineteenth-century engineers' efforts to predict steam engine efficiencies, efforts to improve engineering analysis frequently gave more to pure science than science's discoveries contributed to the creation of new technology.

During the nineteenth century, and even more in the twentieth, the physical sciences made enormous strides forward, so much so that their discoveries became a wellspring for revolutionary technologies and incremental improvements alike. The telegraph sprang from early electrical experiments, and radio from James Clerk Maxwell's equations and electromagnetic wave theory. In the twentieth century, atomic theory and experiment led to the invention of the laser and its many applications that followed. Nuclear energy originated from the observation of fission and the discovery of the neutron, and microelectronics grew out of the understanding of the quantum physics governing electron behavior in semiconductors.

In modern times we find that the venues in which scientists and engineers work are growing closer together, with the methods they employ becoming at times nearly indistinguishable. Indeed, the thrill of the scientist's discovery and the engineer's creation have much in common, and both may bring great intellectual challenge and satisfaction. The success of Francis Crick and James Watson in solving the puzzle of the structure of DNA can be compared with that of William Shockley, Walter Brattain, and John Bardeen in creating the first functional transistor.

Today's engineers are immersed in science. Electromagnetics, thermodynamics, solid mechanics, and more are staples of their education, since scientific considerations permeate even the profession's more routine projects. A vast accumulation of scientific knowledge is embodied in computer simulation packages that enhance the ability to compare and contrast the behavior of design alternatives before they are built. Such knowledge is also essential for the intelligent use of sensitive instrumentation coupled with sophisticated data acquisition software to gain maximum insight from the engineering experiments performed on models, proto-types, and operating systems. Whether engineers' projects are gigantic in scope or microscopic in scale, whether they deal with mass-produced products or one-of-a-kind creations, contemporary practice is permeated with science, and mastery is required of a vast body of scientific knowledge related to man-made artifacts.

However much modern science and engineering are intertwined, though, the two are not the same. The goal of science is to discover the laws of nature and comprehend its behavior, whereas that of engineering is to create technology that serves the needs and desires of humanity. Science deals with what is, comprehending nature as it exists; engineering focuses on the future. It creates new material environments, producing products, processes, and systems that did not previously exist.

The discoveries of basic science tend to be detached, removed from ethical considerations and moral quandaries, as scientists pursue their research wherever nature's laws may take them. But in creating technology, engineers carry ideas and devices away from the research laboratory, and as they do, they encounter a world fraught with compromises, conflicts, and ethical dilemmas. They must participate in decisions as to what ought to be built and how it should be constructed. Their designs must accommodate multiple stakeholders, mediating between the differing motives and conflicting interests of clients, corporate patrons, consumers, governmental authorities, and others. Trade-offs abound, among privacy, conven-ience, and communications advances, as well as among transporta-tion safety, fuel consumption, and environmental degradation. Engi-neers must make compromises in terms of cost, performance, and reliability in accordance with deeply embedded cultural values. They must respond to market pressures and military need, allocating risks and rewards. They are expected to anticipate unintended repercussions of the technology they launch and to ask continually, if failure occurs, who will be harmed, and what will be the consequences?

Conflicting human needs and complex cultural issues compound the technical challenges that engineers face as they devise technology and strive to make it work. But these issues also make the process all the more fascinating. Before construction can begin or manufacturing can commence, two fundamental concerns must be addressed: what is to be made, and how it is to be built? The process of answering these questions may begin with negotiations with a corporation, municipality, or other client concerning the characteristics of a unique artifact: a supertanker, a dam, or an oil refinery. Conversely, it may be initiated through market research to determine what the public will buy, in what volume, and at what price. It may involve redesigning, refining, or advancing earlier products in anticipation of shifting consumer tastes or of demand for increased performance or reduced cost. It may be centered about the introduction of entirely new technology whose value must be demonstrated. The idea of cooking with microwaves, for example, had to be introduced along with the ovens themselves, just as the acceptance of using lasers for eye surgery had to go hand-in-hand with the introduction of the equipment itself.

Engineers proceed by formulating design concepts and examining alternative approaches to meeting their clients' needs. What should the architectural layout of the product be, and what should be the principles of operation, sources of power, and materials of construction? Rarely will engineers have anything approaching a clean slate in considering such alternatives. Their firm's history, expertise, and existing manufacturing capabilities will constrain their options to a greater or lesser extent, just as the political climate and cultural norms will limit the range of what they can consider. Nevertheless, all the latitude available must be used to probe deeply into which combinations of principles and configurations can come closest to meeting all of the often conflicting requirements for performance, dependability, and price. If they cannot all be met or exceeded, how are the compromises to be made that best serve client interests or market demand? To what extent can new or innovative technology be introduced? Has scientific understanding reached the point where its findings can be reliably implemented with the resources at hand and within the time available?

The conceptual design arising out of discussions, sketches, and debate must be converted to detailed drawings and specifications from which the technological product can be built. Electrical engineers use computer-aided design packages to lay out the geometric intricacies of circuitry that will go on a chip. Structural engineers manipulate beams, struts, and columns by computer simulation as they develop their designs. Mechanical, chemical, and other engineers likewise rely heavily on computers as they resolve issues at increasing levels of detail. As they proceed, the engineers must not only determine what is to be built but also the sequence of steps needed to build it. The process by which a computer chip is built up layer by layer affects the design of the circuitry, just as the sequence of construction required to span a river dictates the form that a bridge may take.

Creative challenges go beyond the sophisticated computer tools used in analysis and design. They arise in knowing the limitations of those tools and in having respect for what they are unable to predict. Engineers must understand the physical behavior of the materials and mechanisms that may not be captured in computer simulations. Regardless of the level of an engineer's experience, the sense of anticipation and nervousness that some serious design flaw has been not been recognized is ever present for the designers. It grows as concepts are detailed in drawings, and drawings are realized in hardware.

Whether a computer chip or an automobile, a home appliance or a jet airliner, excitement mounts as drawings are implemented in a physical prototype. Parts and components, particularly those that are new or complex, are tested, modified, and refined early on, before risking their incorporation into the design of a larger system. Computer models assist in the decision-making process. They simulate the prototype's behavior, predict weaknesses, and allow shortcomings to be eliminated before paper becomes reality. But then the day of reckoning comes: the prototype is ready to run.

Questions are present in each engineer's mind: will the prototype meet the design criteria? Will it even work? At this point the old adage known as Murphy's Law best summarizes the collective mental state: "If anything can go wrong, it will." Silent wishes abound: wishes that there had been more time or a larger budget for more extensive component and part testing; that more powerful computer simulations could have been performed; and that scientific understanding was more complete. The imperfect state of knowledge and limited ability to predict performance is foremost in everyone's thoughts. This is the engineers' anticipation and anxiety when a newly designed computer chip is powered up for the first time, a prototype automobile rolls out onto the proving ground, or a test pilot taxis an aircraft toward its first flight.

The exhilaration is palpable when the prototype performs well. But exaltation is usually short-lived, for surprises inevitably turn up between drawings and hardware. The prototype tests make apparent the oversimplifications in the computer models, the inadequacies of the parts' tests, and the inability of the designers to anticipate every nuance of behavior. Thus, still to come are extensive periods to test, fix, and redesign, as unforgiving physical reality reveals design flaws and construction shortcomings. Many tests lie ahead to assure adequate performance under adverse environmental conditions and to confirm the design's durability. Still, also to come, is completion of the tooling, machinery, and plant design for mass production; these will loom large in determining the product's quality and the cost at which it can be produced. A great deal must transpire to assure that laudable prototype performance is translated into an acceptable mass-produced product. Can it be assured, for example, that the test-track performance of the hand-built prototype will be duplicated by every mass-produced auto, even if it is ten years old, maintained by a novice, and operated on slippery roads in subzero weather?

The methods that engineers employ must vary with the technologies they create. The engineering of consumer products, where visual image, pleasant touch, and convenience in use are major selling points, calls for greater attention to artistic industrial design than does the engineering of manufacturing machinery, where these characteristics are secondary to efficiency, cost, and maintainability. Architectural aesthetics must play a much larger role in the design of a landmark skyscraper than of a chemical processing plant. The development of computer software entails unique challenges that depart significantly from those encountered in the design of the hardware on which it is to run. In creating one-of-a-kind technology, engineers face situations not encountered with mass-produced products, for there are no prototypes to test before building a skyscraper, supertanker, or suspension bridge.

Engineering takes place at locations as diverse as the technologies that it creates. At construction sites, massive structures rise from the earth amid beehives of equipment and human activity. Progress is often a battle with wind, rain, mud, or whatever other obstacles the weather may bring. In contrast, semiconductor fabrication plants concentrate billion-dollar complexes of automated equipment in superclean environments. There complexity is found on a microscopic scale, with millions of transistors etched on single wafers of silicon, the details too small to be seen through ordinary microscopes, and where tiny dust particles can be as damaging as a crane collapse on a construction site. Nevertheless, amid such diversity in design and production, underlying themes unite the engineering profession.

Across a wide spectrum—between gigantic and microscopic extremes, from mass-produced products to one-of-a-kind projects—engineers pursue the creation of the products and systems that make our world. It is a unique profession, made more difficult to understand by the diversity of its endeavors and shaped by the responsibilities that society places upon it. Indeed, identifying the underlying themes that cut across all engineering seems more difficult than relating the specialized practices in medicine or law to the whole of those professions. But the divergence between engineering, medicine, and law goes deeper than that.

Physicians combat illness to increase longevity and quality of life. They may be faulted for poor judgment or sued for negligence or incompetence, but they are not held responsible for creating the ailments they must heal. Similarly, lawyers are charged with protecting the rights of the citizenry and resolving conflicts in a just and equitable manner. Attorneys may be criticized for conflict of interest, incompetence, or worse. Yet they are rarely held responsible for creating the disputes or fomenting the antisocial behaviors that give rise to much of their work.

Engineering, however, is different. Unlike scientists, who seek comprehension of the natural universe, or the professions that strive to ameliorate existing problems—restoring clients' health or adjudicating their conflicts—engineering strives to create technology and to make it work. Its focus is on opening new opportunities that, if approached wisely, may produce great benefits. But new technologies bring about change, and change can be disruptive as well as rewarding. Unforeseen side effects may appear, and technologies may fail, sometimes disastrously. Doctors and lawyers must live with the knowledge that a single mistake may be fatal to a patient or send a client to prison or worse. But more haunting still, the designers of aircraft, automobiles, buildings, and bridges must deal with the reality that a single design flaw or production defect, when gone undetected or uncorrected, may send many to their deaths.

What, then, lies behind this profession that has played such a central role in creating the technological world of today? The essence of engineering lies deeper than the tools of contemporary practice and transcends the current state of scientific knowledge. Though deeply intertwined with science, engineering is also a social enterprise, and examining the historical roots from which it evolved is invaluable in deciphering it nature. Its development is a fascinating tale in its own right, one that is pertinent not only to engineers, but to all citizens of contemporary society. Wise public policy and informed consumer choice rests on an understanding of technology and of the enterprise through which it is created. Just as the quest for genealogical knowledge of our personal roots aids in understanding who we are, examining the roots of our technology and those who created it may shed light on how we came to be the civilization that we are.

The Practical Arts of Our Ancestors

I n this industrial age of automation and mass production, we are
naturally fascinated by ornamental objects of the past. Collections of jewelry, furniture, pottery, and painting produced by artisans of the preindustrial era grow in popularity, as witnessed by the
profusion of antiques shops. Newspaper articles, magazines, and television programs tout the importance of preserving the crafted heritage of our past. But artisans did not design objects that were only
pleasing to the eye before the industrial revolution. Craftsmen in the
"practical arts" provided the artifacts upon which our technological
heritage rests. The goals of these practical artisans were not aesthetic
but functional. Indeed, the functionality achieved in the wagons,
windmills, and waterwheels, in the boats, barrels, and barns that
they produced largely determined the quality of life. And yet,
although such objects were carefully crafted for utilitarian purposes,
they invariably displayed an aesthetic appeal, in which grace and
beauty arose naturally from functional requirements.

We may gain some appreciation for the technological heritage of
those practical arts by examining the artifacts of bygone days and
considering the artisans who worked in the media of stone, wood,
and iron. The handsome pair of farm wagon wheels that my wife and
I purchased to adorn the drive that leads to our vacation hideaway
serves as a splendid example of this technology. Made of durable

woods by a skilled wheelwright, they exude power that radiates outward from their hubs through the spokes to the rims of nearly five
feet in diameter. But in their subtly dished shape, similar to that
shown in figure 2, the slight alternating offsets of the joints between
spoke and hub, and the spokes' outward taper toward the rounded
rim, there lies a graceful beauty that elicits a response somewhat akin
to that felt for fine pottery, furniture, or other, more ornamental artifacts. But the wheels' proportions, the number and positioning of
spokes, the curves and tapers are not ornamental; rather, they evolved
for utilitarian—we might say engineering—reasons.

Our wheels represent a long and venerable tradition, with roots
stretching back to the beginnings of recorded history. The oldest
known wheels were wooden disks, thought to have evolved from the
use of logs as rollers. A Sumerian pictograph from about 3500 BCE
shows a sledge (a strong, heavy sled) equipped with wheels. Potters'
wheels originated in Mesopotamia at about the same time. The
spoke was a seminal invention, for it greatly reduced the wheel's
weight and gave it a resilience that cushioned travel over rough terrain. The first spoked wheels appeared on chariots around 2000 BCE

Figure 2. Jost Amman's woodcut
of a wheelwright's shop, 1568.
(From E. M. Jope, "Vehicles and
Harness," in *A History of
Technology*, ed. Charles Singer, E. J.
Holmyard, A. R. Hall, and Trevor I.
Williams [New York: Oxford Uni-
versity Press, 1956], 2:552.
Reprinted by permission of Oxford
University Press.)

in the Middle East, and they found extensive use in classical Greece and Rome. Wheels played a central role in the medieval development of animal, water, and wind power that ameliorated human toil. They have occupied an essential place in the development of industrial society and remain one of technology's most ubiquitous manifestations.

From truck tires to in-line skate wheels, from jet engine rotors to compact discs, wheels play an integral role in industrial technologies and consumer products. The wheel's rolling motion is central to the operation of gears, cams, pulleys, bearings, and a myriad of other mechanisms upon which we depend. As such, the various modern derivatives of this most ancient of inventions are the frequent result of contemporary engineering activity. And yet if we would like to examine the historical roots of engineering, here is a quintessential artifact. For if the essence of engineering lies in the creation of technology, the wheel offers a perfect example. Wheels were produced in the earliest recesses of recorded history, long before anything that we would classify as science-based engineering existed. Thus, if engineering has dimensions beyond those of the application of science, then from the examination of the wheel, we may begin to understand what they are.

Fortunately, traditional modes of production, which predated most of what we would consider modern manufacturing methods, have coexisted with more modern methods until only relatively recently. Thus it is possible to examine these methods. Our pair of nineteenth-century wagon wheels, produced in rural Wisconsin, may lend some perspective to modern practices in the design and production of technological artifacts in a preindustrial, prescientific age. For the wheelwrights who fashioned these wheels, and the wainwrights who designed the wagons they were to support, adopted practical arts that their immigrant fathers transported across the Atlantic; these methods were rooted in technological traditions that had changed only slowly over many centuries.

Who were the wheelwrights and wainwrights who engineered with wood and iron? They were first of all craftsmen, who learned their trade not through textbooks or formal education but through hands-on experience in making wheels. Although the details vary from locality to locality and generation to generation, just as the wagons they produced evolved, they learned their trade though an apprenticeship tradition that traces its roots to the craft guilds of the Middle Ages. Around the age of thirteen an apprentice would be bound to a master craftsman through an indenture, typically lasting seven years. For the duration of the indenture contract, the apprentice most often lived in home of the master, where he was provided with clothing and other necessities. The apprentice learned the trade by example and direct participation. Beginning with only the most menial tasks and by observing the master, the apprentice's participation in the construction process gradually increased until he was competent in all the craft's aspects.

As the skill of the apprentice grew, the master might provide him with a small wage, and toward the end of the contract he would acquire his own set of tools. Upon successful completion of the indenture, he would be admitted to the local craft guild as a "journeyman," allowing him to journey to other villages and accept employment working under other master craftsmen. Following several years of employment under different master craftsmen, the journeyman might aspire to become a master craftsman so that under guild auspices, he could open his own shop, employ journeymen, and take on his own apprentices. To become a master craftsman, it was often required that the journeyman submit a "masterpiece," from which the guild elders could judge his technical proficiency. In considering the applicant, his social standing and financial assets would be weighed strongly. Needless to say, family ties were important, and frequently craft shops were passed from father to son, to a favored apprentice, or to one who had married a master's widow or daughter.

The scarcity of skilled labor in the New World and the availability of inexpensive land in the west greatly weakened the guild system and often diminished the length and rigor of apprenticeship requirements. Frequently, an apprentice could break his indenture after mastering only the more basic skills of the trade and move west, where the paucity of skilled tradesman would allow him set up shop without the certification of guild membership. Nevertheless, the design and construction techniques employed in our Wisconsin wagon wheels, as in most other preindustrial technology of that era, were shaped by the ancestral traditions brought by immigrant craftsmen to our shores.

And what were the skills inculcated by this guild tradition? How did the wheelwright go about making our wheels? With the simplest of hand tools, he acquired expertise in working wood, in shaping the hub and spokes, and in forming the wheel's rim from curved segments of wood called felloes, as shown in figure 3. The texture and grain of the wood had to be carefully considered to assure adequate strength; the hint of a knot or other irregularity could destroy the integrity of the entire piece.

After shaping the hub, the wheelwright would skillfully chisel the squared holes, called mortises, where each of the spokes was to be inserted. Likewise, he had to shape carefully each of the spokes and felloes. Placing the hub across a narrow trench called a wheelwright's pit, he would use a sledgehammer to pound each spoke into the hub,

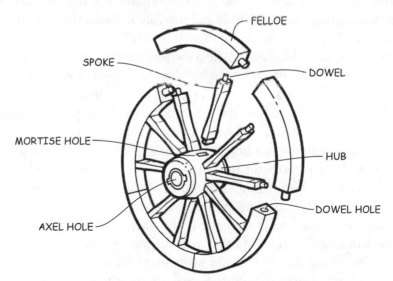

Figure 3. Components of a wheel. (Diagram by James Hand.)

at the same time taking care not to crack it. Removed from the pit, the wheel was then "ringed" by forcing the felloes onto the spokes to form the rim. After the wheel was formed and balanced, the wheelwright would place it flat on the ground and bore the central hole through the hub where the axle would later go. Through the entire process, constant centering and checking for accuracy of the angles at which the spokes met the hub and the rim met the spokes were key to the successful construction of a balanced wheel. All this was accomplished without bolts, screws, or glue. Only the tight fit of the spokes in the mortise holes secured them the to hub and felloes, and only wooden dowels aligned felloe to felloe to form the rim.

Success in forming and assembling the wheel's component parts depended upon the wheelwright's profound understanding of wood and what we would now call "materials selection." Each of the wheel's components called for woods with specific properties: ash, beech, or elm would be preferred for the rim, because they combined the flexibility required in fabrication with the toughness demanded by the rugged use to which the wheel would be put; oak was the choice for the spokes, because its strength allowed for great weight to be supported on those slender shafts; and elm was selected for the central hub because it would best resist splitting when the wheelwright chiseled square mortise holes into it and forcibly inserted the spokes.

The tree's species was only one of several important factors in choosing the wood for each wheel's components. The experienced wheelwright would often supervise the cutting of the timber. He would note the natural curves and crooks of standing timber, cut it, and then leave it for an extended period to become properly aged, sometimes, as in the case of oak, for up to four years. The wood's qualities also varied according to the soil and climate where the tree grew as well as the season in which it was cut. Timber had to be properly dried to be tough and durable. Moisture content had to be limited to reduce weight and avoid the dark borings of maggots. The workability of the wood depended not only on the aging period: temperature, moisture, and other conditions of exposure were crucial factors both in how it could be bent, cut, and shaped into the wheel and how it would endure the rigors of usage.

The final steps in the creation of the wheel required the skill and cooperation of the village blacksmith. The assembly of the wooden parts that constitute the wheel gather strength and protection from an iron tire. Made red hot in a circular fire, this hoop was placed around the wheel's wooden rim and was quickly doused with water before the wood could start to burn. The cold contracted the tire, compressing rim onto spokes and spokes into hub to form a unified structural whole. Precision was crucial, for in rugged use too loose a tire would run off the rim, while too tight a tire would distort or break the wheel. The blacksmith then joined with the wheelwright in inserting an iron sleeve into the hole that had been previously bored in the hub to provide the bearing surface for the axle. In our wagon wheels, shrinking smaller iron bands around either end of the hubs to further increase their strength completed their construction.

The building of wheels and of wagons, of course, was closely related. Since the wheel was the part of the wagon whose construction demanded the highest level of skill, a master wheelwright frequently also became the village wainwright, or wagon builder. In an established shop the master craftsman typically employed several journeymen and also taught the trade to young apprentices. The shop was invariably a family venture; the master's wife was expected to look after the needs of the young apprentices and often took care of the bookkeeping as well.

Each project would begin as a farmer discussed with the village wright his needs and tastes, and what he liked or didn't like about the wagons he had owned. Face to face they would negotiate the cost and character of the wagon to be built. The shop proprietor had to possess an ingrained sense of how to adapt his ancient craft to produce wagons and wheels attuned to the local terrain, climate, and patterns of use. And as a result, no two wagons were exactly alike.

For success, a wagon's chassis and wheels had to form an integral whole, as shown in figure 4. Each pair of wheels required careful mounting and balancing; otherwise the wagon could wobble or tilt. Large-diameter wheels were preferable because they reduced the force required to pull the wagon, but smaller-diameter wheels were required on the front axle so that turns could be made in a reason-

Figure 4. Wisconsin Farm Wagon. (From a watercolor by Phil Austin, A.W.S.)

ably small radius without causing the wheel to rub against the wagon's side. Conscientious design trade-offs between desirable characteristics, however, did not decide the wheel's diameters; quite to the contrary, they subtly evolved from years of experience and generations of wagon use.

Completion of a new wagon did not end the wright's labors. Skill in construction was nearly inseparable from long experience with repair. Much of an artisan's work, whether that of wheel- or wainwrights, consisted of repair as well as new construction. During seasons in which use was heaviest, repair took first priority, with new construction relegated to times when breakdowns were less frequent. In the wainwight's shop or the farmer's field, where the hub had cracked, a spoke had broken, or a tire had come loose, repair was ever necessary. While making repairs, the journeyman listened to what the farmer did or did not like about the wagon and what he would want modified, should he buy another. This feedback was typical of all crafts in preindustrial times, whether the artifact was a wagon, a windmill, a boat, or a building.

Repair played an integral role in the evolution of craft products, because the craftsman gained an intimate knowledge of how failures occurred and where the weak points were. This sense of strengths and weaknesses was expressed in small modifications and incremental changes to correct deficiencies and enhance the performance of each new wagon to emerge from the shop. Where breakage occurred, more material was retained, but where it did not, more wood could be shaved to save weight. Proportions might change slightly, dowels made longer, or tires shrunk more tightly. In modern parlance we would consider this to be a form of "design optimization." But it was less self-conscious than that; it came out of the knowledge gained through continual experience in how the wagons were used, the conditions they faced, and how they failed.

The village shops produced wagons with wheels that were elegant in appearance and finely attuned to agrarian needs. The Wisconsin wheelwright crafted our wheels with characteristics that evolved from those he or his immigrant ancestors had brought to the New World. He likely had no explanations satisfactory to a present-day engineer for why these characteristics were desirable, since his understanding of wheel design was intuitive, ingrained through apprenticeship and years of practice. His strength was his detailed fingertip knowledge of how wheels were to be constructed and adapted to the local farmers' needs. Nevertheless, each of the subtle features that enhance our wagon wheels' appearance also provided utilitarian benefits. The benefits that the wheelwright understood intuitively now may be demonstrated analytically through stress analysis, fracture mechanics, and other techniques of modern engineering.

The slight dishing or saucerlike characteristic of wagon wheels dates back at least to the dished wagon wheels unearthed from the first century BCE. A wheel's dishing—with the top of the dish facing outward from the wagon—can in fact be shown to increase the wheel's strength toward lateral, or side-to-side, forces without added material or weight. This strength was crucial in resisting the effects of the periodic rocking motion imparted by each step of the horse as it plodded on its way. If the wheel were perfectly flat, the side-to-side motion would tend to bend the spokes. But with dishing, the bending is eliminated and the forces become simple pushes and pulls along the spoke's axis. Just as it is easy to break a toothpick by bending it, but nearly impossible to do so by pushing or pulling on the ends, the spokes of a dished wheel could resist the side-to-side heaving of a heavy wagon but would break if the wheel were flat.

To be effective, the dishing had to be taken into account during the construction of the wooden axle. The axle had a slightly concave downward bend, causing the wheel to slant a bit outward at the top. The slant was just enough that, at any time, the spoke that was instantaneously carrying the weight of the wagon—the spoke between the wheel's hub and the point on the rim in contact with the ground—was nearly vertical. The vertical spoke was able to withstand the maximum compressive weight of the wagon. Conversely, the opposite spoke—the one pointed skyward—slanted outward, which was permissible since at that instant it carried no weight. At the same time, the outward slant of the upper half of the wheel had other advantages. It allowed the wagon body to be tapered outward, increasing its load-carrying capacity without lengthening the axle. In fact, the axle could not be increased beyond the standard length, since the wagon had to ride in the ruts of the local roads. Finally, the outward slant prevented the muddy rim from passing directly above the hub and dripping mud onto the axle bearing.

In addition to dishing, which has been a feature of wheels for millennia, the appearance of our nineteenth-century wheels is further textured by an elaboration of dishing that evolved within the last few hundred years. Every other of the fourteen spokes is attached to the hub at a slight offset from its two neighbors. This staggered spoke configuration, which is seen in a more pronounced form in modern bicycle wheels, has several benefits. It gives the wheel the shape of not one dish but two, with the outer being the more shallow and facing the inner. Though difficult to picture, such spoke staggering adds rigidity to the wheel, which exceeds that of a singly dished wheel in resisting lateral motion. This double dishing also decreases the need for bending the axle downward. Furthermore, if the mortises, where each of the spokes enters the hub, were not staggered, the stress in the hub would be amplified in the proximity of the corners of neighboring mortise holes. Thus, by offsetting every other spoke and mortise, the stress concentration is alleviated. This in turn decreases the likelihood of the hub splitting at the mortise holes. The staggered spoke placement results in a stronger, more durable wheel and allows a smaller and therefore lighter hub to be used.

Much of the detail in the curvatures of the hub, spoke, and rim, which contributes to our wagon wheels' graceful appearance, results from the extensive shaving performed by the wheelwright. Shaving removed excess weight from the wheel and thereby eased the horses' effort in pulling the wagon. The skilled artisan pared down timber and hewed away wood wherever his tacit knowledge told him that stresses should be small. Wherever strength was needed, where stress was concentrated around a hole or mortise, or where heavy loads bore down, nothing was shaved away. This was true for the wagon as well as for the wheels. As much as an eighth of a wagon's weight might be shaved away while preserving the essential strength of the structure.[1] The wagon's sides were tapered down where less strength was required, and the tapered wagon, like its wheels, appeared more graceful as a result. Craftsmen might argue that such a graceful appearance was not deliberately sought, but they knew it looked right according to the tradition passed from generations of forebears and according to long experience in construction and repair.

Over generations, incremental improvement resulted in artifacts that were exceedingly well suited for their intended use. Within this tradition, however, innovation posed problems and thus was very slow to occur. Even the most ingenious craftsman faced nearly insurmountable barriers in any attempt at improvement that substantially altered the design. Suppose, for example, that our wheelwright wanted to make more radical improvements: perhaps to reduce the cost and effort of building the wheel, or maybe to allow it to support a heavier wagon, or to make the design more dependable or durable. Instead of just varying the dimensions slightly, or shaving a bit more or less of wood from various places, he might like to redesign the wheel with a reduced number of spokes. This would certainly reduce the number of mortises to be cut and the number of spokes to be shaped, and might therefore result in a wheel that was more economical to build. But if he were to set out on such a path, he would inevitably be confronted with three closely related classes of problems.

First, he would need some method for estimating how thick to make each spoke if he were to reduce the number. But this was problematic, for he had no quantitative design rules available to tell him how the stress levels in the spokes related to their number and thickness for a fully loaded wagon. In fact, the situation would be worse than that. For even if he had a means of estimating the stress, he could not proceed without data on the stress level at which the oak would be expected to break. While such data may have existed in the nineteenth century, he would not have had access to it. Neither would he be able to determine a reasonable factor of safety, a ratio of the breaking stress to the actual stress that would allow for the variability in the properties of the wood and for the simplifications made in his calculations. The practical arts that he practiced did not equip him to carry out such an analysis. The lore of the craft by which his wheels were built was incapable of predicting the strength of the untested wheel, for he had no access to engineering science— the science of manmade artifacts.

Under these circumstances the only road open would be the riskier one of simply guessing a reasonable thickness for the reduced number of spokes. Even if the guess turned out to be an excellent choice, and the resulting configuration yielded a substantial saving in the oak needed and therefore in cost, the wheelwright would risk new difficulties in construction. Fewer spokes might require that the number of felloes making up the rim be reduced, with each felloe being longer. Would that make them more difficult to fabricate? More importantly, would the ringing process, in which the felloes are forcefully fit onto the spokes and aligned to each other, become more difficult? If the wheelwright were to encounter such problems, he could lose days of his precious time. Other income-producing activities would be put aside while he adjusted his craft construction techniques through trial and error to accommodate the new spoke design.

Let us suppose that a set of wheels of the new design could be built successfully and installed on a wagon. The wheelwright would then be faced with a second set of problems related to design evaluation. How is he to know that the new wheels are better than the old? He may already have some sense of relative cost based on the amount of oak required in constructing the spokes and the labor required to assemble the wheels, but evaluating their performance would be a more difficult matter. His customers, the local farmers, would likely be leery of an obvious change from the traditional wagon and might refuse to risk their money to provide field trials for the new design. Even if some more daring souls were to cooperate, the wheelwright would be putting his reputation on the line, and, more importantly, he would be responsible for repair or replacement should failures occur. If the new wheels did seem to be working out, and the farmers felt that they were an improvement, the positive feedback to the wheelwright would most likely be erratic and slow in coming—too slow to assist in deciding whether to risk building his next wheels with the new design or to revert back to the traditional number of spokes.

The wheelwright needed tests that he could perform before the farmer put the wagon into service. Such tests would serve to verify that the design actually was an improvement and/or to identify weaknesses, so that problems in the new configuration could be fixed before the wagon was turned over to the farmer. Tests would be indispensable in ascertaining whether the wagon could support a heavier load and whether it could be pulled with less effort. Evaluating design issues such as whether the wheels would be as durable as the traditional forms and how well they would behave under adverse weather conditions or difficult terrain would demand experimental resolution. Such evaluations constitute a major part of modern engineering practice. But even if the wheelwright had adequate time and financial resources, such testing was inconceivable. Performance testing requires methodologies that were beyond an artisan's experience and requires test instrumentation that didn't exist in the wheelwright's world.

Beyond the lack of predictive design rules and of verification test methods, a third, more pervasive factor precluded craft workers from undertaking more radical attempts at technological innovation: a lack of financial resources. Any attempt to alter or improve significantly an existing technology entails risks. Design knowledge is gained through the iterative process of design, build, test, and redesign. In this process, resources are expended and time is consumed with little assurance of success. For the preponderance of craftsmen, regardless of trade, and throughout the preindustrial world, the likelihood of failure with the resulting loss of time and resources made the risks of such a process too great to contemplate. Artisans typically operated from small shops, holding minimal financial reserves. Valuable raw materials could not be wasted, and time could not be devoted to building an artifact that might turn out to be unusable. To assure compensation, wagons had to be built to conform to the immediate expectations of the prospective buyer. Unacceptable risks also loomed in costly repair efforts. Moreover, denigration of a craftsman's reputation would result, should unanticipated weaknesses appear in the innovative design. Thus, without a patron whose deep pockets could afford the risks of setbacks and failures, the concerted effort required for innovative new designs would not be forthcoming. Design evolution could come only through small, nearly imperceptible alterations, arising naturally in the making of artifacts. Economically, the wainwright's shop was confined to build wheel after wheel and wagon after wagon with only those slight modifications that were acceptable in adapting each wagon to the needs and tastes of the tradition-bound farmers who bought them.

Our wheelwright's dilemmas illustrate the difficulties in pursuing technological innovation before the development of the intellectual tools and economic resources now available to engineers of the industrialized world. Within this framework, the slow evolution of technology among the traditional crafts is quite understandable. At another level, these obstacles to progress make the refined functionality and elegance displayed by the abundance of artifacts arising from craft tradition all the more amazing. The characteristics of craft technologies have inspired scholars to draw a number of analogies between technological evolution and that of living organisms.[2] Provided we do not stretch such analogies too far, they serve to broaden our perspective on our technological roots.

According to Charles Darwin, species evolve through the process of natural selection and by "survival of the fittest." Of the small variations among individuals of a species, some favor survival, and on average the individuals possessing these favorable variations live to pass them on to more offspring—that is the definition of natural selection. Thus the species accumulates favorable deviations or changes, and conversely, unfavorable variations are bred away. In this way a species gradually adapts itself over many generations to its changing environment.

Similarly, over time, artifacts were crafted to suit the particular needs of their owners' situations. Changes in artifactual characteristics compensated for known weaknesses and were adjusted to the changing availability of materials, catering to the uses and tastes of individual customers. Each change, however, tended to be small, as craftsmen shunned the risk of failure inherent in venturing too far from the tried and true. From village to village across many localities a profusion of such changes would appear over time, some as deliberate trial, but most purely by chance. As a result, many variants of wheel or wagon design coexisted at any time. The intricacies of the wagons varied as much as the soil, the climate, and the crops of the farmers who bought them. This natural variability within the conservative craft tradition provided a mechanism for a form of natural selection. Whether designs were modified by accident or purposefully, users would develop preferences for one wagon or another based on their sense of what best filled their needs or what caused them the least difficulty. These preferences then influenced the craftsman when negotiations began for the next wagon to be built.

Biological evolution proceeded from simplest forms of life to more complex ones. Organisms evolved from lower to higher forms. Beginning with single-cell organisms, cell differentiation took place, leading to separations of function. Invertebrates evolved to creatures with backbones and increasingly complex skeletal forms, thus separating structure from other functions. Likewise, digestive systems evolved to facilitate varied forms of nutrition. And sophisticated temperature control mechanisms developed, allowing warm-blooded animals to survive in more varied climates.

Wheels, likewise, evolved from the simple wooden disks that were the first wheels. To go back even further, the wheel's origin can be traced to the rollers used to assist in sliding goods on sledges. Later, the disks were superseded by wheels with components: the hub, spokes, and rim. Each served a specific function, and each was formed from the wood best suited to meet that function. Further development took place with the advent of the iron tire, which protected the wood from wear and served to compress the rim on the spokes, thereby adding strength. The insertion of the iron sleeve between wheel and axle created a bearing surface that decreased friction and ameliorated wear. Some of these improvements may seem quite drastic, but they were spread over many centuries and were superimposed on the profusion of smaller changes in dimension, proportion, and material that made up the historical evolution of the wagon wheel.

Following the publication of Darwin's thesis, his critics questioned its ability to explain the beauty of the plant and animal world. Darwin's opponents argued that those features must have an aesthetic basis and not result merely from natural selection. With further study it became more and more widely accepted that even the beauty of nature could be explained in terms of this theory. A botanist, for example, can justify bright colors and graceful shapes and smells of wild flowers in terms of their ability to attract the insects needed for pollination. The argument is not dissimilar to the one made here that the graceful lines of the wheel can be explained in terms of functional requirements. Aesthetic appeal can have its roots in the economical use of energy and material, of efficiently proportioned parts, and of reduced complexity that evolved for functional purposes.

Without books to learn from, scientific principles to guide them, or drawings to rely on, artisans—our technological predecessors—created graceful and efficient artifacts ideally suited to their practical purposes. Yet as primitive as their methods may appear to those of us grounded in the scientific methodologies behind today's technology, the heritage of generations of craft workers survives in the design practices of today. Much of that heritage is embodied light-heartedly in the lines of a poem by Oliver Wendell Holmes (1809–94) titled "The Deacon's Masterpiece." It begins,

> Have you heard of the wonderful one-hoss shay,
> that was built in such a logical way
> It ran a hundred years to a day,
> And then of a sudden, it—ah, but stay,
> I'll tell you what happened without delay,
> Scaring the parson into fits,
> Frightening people out of their wits—
> Have you ever heard of that, I say?

In a later stanza, Holmes insightfully details some of the difficulties encountered in creating an artifact with a hundred-year design life:

> Now in the building of chaises, I tell you what,
> There is always, somewhere, a weakest spot,—
> In hub, tire, felloe, in spring or thill,
> In panel, or crossbar, or floor, or sill,
> In screw, bolt, thoroughbrace,— lurking, still,
> Find it somewhere you must and will,—
> Above or below; or within or without,—
> And that's the reason, beyond a doubt,
> That a chaise breaks down, but doesn't wear out.

Although the nomenclature of nineteenth-century carriage components may not be readily recognizable, it's apparent that the design
and manufacturing challenges of eliminating failure modes and
weakest links were as much on the mind of Holmes's imagined
wainwright as they are in the thoughts of today's engineers. With
enough literary talent one might attempt substituting engine parts,
circuitry, hydraulic lines, and other components of a computer, aircraft, or automobile to pen a modern parody.

The carriage, depicted in figure 5, was not only a success, but
toward the end of its hundred-year design life it met the ideal that
many engineering design teams pursue, but few ever approach:

> There are traces of age in the one-hoss shay,
> A general flavor of mild decay,
> But nothing local, as one may say.
> There couldn't be,—for the Deacon's art
> Had made it so like in every part
> That there wasn't a chance for one to start.
> For the wheels were just as strong as the thills,
> And the floor was just as strong as the sills,
> And the panels just as strong as the floor,
> And the whipple-tree neither less nor more,
> And the back crossbar as strong as the fore,
> And spring and axle and hub encore.
> And yet, as a whole, it is past a doubt
> In another hour it will be worn out!

And one hundred years to the hour after the wagon entered service,
the Deacon's successor of several generations experienced the end of
the design life:

> First a shiver, and then a thrill,
> Then something decidedly like a spill,—
> And the parson was sitting upon a rock,
> At half past nine by the meet'n-house clock.

And all the wagon's parts indeed had not only equal strength but
also equal durability:

What do you think that the parson found,
When he got up and stared around?
The poor old chaise in a heap or mound,
As if it had been to the mill and ground!

You see, of course, if you're not a dunce,
How it went to pieces all at once,—
All at once, and nothing first,—
Just as bubbles do when they burst.

End of the wonderful one-hoss shay.
Logic is logic. That's all I say.[3]

Although the poem was written well over a century ago, it captures design ideals that endure from the craft traditions of preindustrial times to the high-tech engineering of today. Critics said—although Holmes denied it—that the poem, in its completeness, was a metaphorical commentary on the Calvinist theology of the time.

Figure 5. The one-hoss shay, as drawn by Howard Pyle with the caption "A general flavor of mild decay." (From Oliver Wendell Holmes, *One-Hoss Shay* [Boston: Houghton, Mifflin, 1891].)

Still, the narrative holds. We may chuckle at the possibility of constructing anything so perfectly that nothing breaks and everything
wears out at exactly the end of its design life. Theological implications aside, the portrayal of craft values in "The Deacon's Masterpiece" serves as a reminder to present-day engineers that although
our profession relies heavily on the methods and findings of
modern science, it has much longer-standing roots. It is the benefactor of venerable craft traditions and the recipient of the age-old
creative drive of man, the maker.

Three

Maximum Height and Maximum Light

The towering tones of the brass rising above the orchestra and organ in the great fugue that is the finale of the Saint-Saëns Third Symphony always transplant my thoughts to the first trip that my wife and I made to France many years ago and to where we first experienced the grandeur of medieval engineering. One morning late in June, we took down our tent at a Paris campsite along the Seine, packed all of our gear into our Volkswagen, and headed out of the city. We traveled southwest across fields of undulating grain. The landscape was similar in many respects to the plains of the midwestern United States, where we had grown up. But then in the distance a pair of spires appeared on the horizon above the expanses of grain and reached ever higher as we approached, until in full view arose the magnificent Gothic cathedral above the ancient town of Chartres.

With the cathedral before us, I found humming the rising strains of the Saint-Saëns symphony irresistible as our eyes moved upward with the sweep of the cathedral's stone structure, shown in figure 6. Deep below the cobblestone pavement the foundation encompassed the crypts of the more ancient Romanesque churches, dating from the fourth century, when the earliest Christians arrived in this region. Looking upward we marveled at the massive stone statuary surrounding the doors, perched above them, and looking downward from pinnacles. Still higher, our eyes rested on the profusion of figures displayed in the stained glass of the upward-pointing Gothic windows. Even then, we were seeing only a fraction of the more than ten thousand figures in stone and glass that adorn the edifice. Sweeping upward between the windows, the elaborate double tiers of flying buttresses provided strength and structural stability but at the same time appeared light and filigreed, with the openness of their arches silhouetted against the afternoon sky. Yet higher, above the stained glass and arch work, the steep gabled roof rose to meet the two spires that reached ever higher toward the heavens. We mar-

Figure 6. West front, Our Lady of Chartres cathedral. (Photograph by A. F. Kersting, in Kenneth Clark, *Civilisation* [New York: Harper & Row, 1969], p. 51. Reproduced by permission of Anthony Kersting.)

veled at the enduring heights reached by the medieval builders with only stone and glass, heights not be equaled for nearly seven hundred years, not until engineers of the nineteenth century built with iron and concrete.

The cathedral before us was the culmination of a succession of more ancient churches. The massive wooden rafters, ceilings, and roofs of the earlier structures made them susceptible to fire; conflagrations in 743, 858, and 1194 destroyed the cathedral's predecessors. But Our Lady of Chartres has now stood for nearly eight centuries, thanks to the innovative stone construction pioneered by the medieval cathedral builders. It survives because, unlike its predecessors, timber construction is confined to the steep gable of the external roof. The gable forms the outer layer of a double roof system, whereas the stone vaulting of the sanctuary ceiling forms the inner layer. The timber gable, which had been covered with slate or lead sheeting, provides weatherproofing for the leaky and possibly cracked vaulting below, whereas the stone vaulting prevents roof fires, ignited by lightning or other causes, from propagating downward and destroying the sanctuary.

On entering the cathedral we were struck by the height and refinement of the vaulted ceiling, as shown in figure 7. It seemed virtually to float above the sanctuary, supported only through the slender columns of stone that connect it to the earth. The columns framed the vast expanses of stained glass that bathed the interior with jewel-like light, glass famed throughout the world, particularly for the Chartres blue. The windows portrayed biblical characters both major and minor as well as caricatures of kings, nobles, merchants, and craftsmen. What a powerful impact this edifice must have had on the peasants, tradesmen, and pilgrims as they came from the crude dwellings of the Middle Ages to enter and celebrate the great festivals in solemn splendor. In such surroundings the medieval hailing of light as the manifestation of the divine becomes more understandable. Instantly I recalled architectural historian Robert Fitchem's characterization of the aesthetic goal for these Gothic structures as "maximum height and maximum light."[1]

While struck by the aesthetic impact of the interior, I began to

Figure 7. Interior, Our Lady of Chartres cathedral. (Photograph by A. F. Kersting, in Clark, *Civilisation*, p. 57. Reproduced by permission of Anthony Kersting.)

think of the engineering feats culminating in the majestic edifice before us. A host of questions presented themselves. What were the human needs or desires to be served, and how were they translated into design criteria that the builders could strive to meet? Who were the architect-engineers responsible for these achievements, and how did they combine ingenuity, knowledge, and experience to design this durable structure to satisfy those criteria? How did they erect the physical embodiment of their vision within the limits of the construction technology and workforce skills available at the time? And finally, how was the effort organized, and who had the patience and financial resources to see it to fruition?

Such questions, of course, present themselves for modern-day technology, whether an aircraft, a skyscraper, or a washing machine. Yet they are still more fascinating when applied to the structures engineered before the scientific or industrial revolutions. Their answers tell us something of the emergence of engineering from small-scale craft production to grappling with the challenges of projects of monumental proportions. Moreover, many of the issues that were resolved in the creation of cathedrals, sailing ships, and windmills transcend the medieval state of scientific understanding and industrial know-how to endure in the practice of engineering today.

Maximum height and maximum light were structural design criteria as well as aesthetic goals for the architect-engineers of Chartres. The criteria were structural answers to a set of religious and civic needs arising from the geography of northern France and the economic conditions of the late Middle Ages.

Unlike the pagan religions of antiquity, in which only priests were admitted to holy sanctuaries, Christianity of the Middle Ages sought to accommodate the largest possible number of the faithful. These medieval edifices were the community meeting place, where the people participated in the ceremonies of the frequent feast days, where illiterate masses absorbed religious inspiration from the light streaming though the biblical stories portrayed in stained glass. Regular civic meetings took place within the cathedrals to such an extent that some towns found it unnecessary to have a town hall. Graduation ceremonies, lawsuits, and even everyday business were conducted there. The sanctuaries strove to accommodate not only the locals but also those making pilgrimages from afar. At Chartres sacred relics said to belong to the Virgin Mary attracted throngs of pilgrims. A primary design criterion must have been for a very large area of open floor space, unobstructed by interior pillars, posts, or walls. And floor space they achieved; Gothic sanctuaries are immense. Chartres' interior space is larger than a football field. Some Gothic cathedrals contained more than enough floor space to accommodate the entire population of the town in which they were situated.

Unobstructed floor area without commensurate height, however, would have had dubious aesthetic impact. It would not create the sensation of reaching to the heavens that the church fathers desired. To achieve that impact the cathedral builders strove to reach to heights in ever-greater proportion to floor area. The heights achieved by the medieval architect-engineers were startling: the vaulting at Chartres, for example, is nearly 120 feet above the floor. A ten-story building could be erected within the sanctuary with height to spare. The spires reach much higher, to nearly 350 feet, as tall as a thirty-five-story building! In terms of light as well as height Our Lady of Chartres is an amazing feat. Between the stone columns that support the vaulted ceiling and roof, over twenty thousand square feet of stained glass covers the vast window openings. These expanses resulted from the builders' unceasing efforts to widen the openings in the walls without compromising the cathedral's structural integrity.

The emphasis on light, even more than on height, represented a shift in design criteria from the early days of Christianity. The earliest churches, those of the late Roman era, were built primarily in Mediterranean climates, where the incentive was to protect the interior and its fresco-covered walls from the blistering sun. Thus builders favored thick walls with few small openings. As time passed, the emphasis on space, and more so on height, grew stronger and gave rise to the Romanesque cathedral, which reached for the heavens. Limited light, however, entered the interiors of these structures, for the massive columns and walls needed for support curtailed the size and number of windows that could be included in their design.

The challenge of obtaining light as well as height was paramount for the designers of Chartres and the other cathedrals of the later Middle Ages. Their structures were to be located in the emerging cites of northern Europe, where the sun was less intense and the weather cloudier. Thus light was welcomed. The biblical stories that had been taught earlier by frescoes, painted on the walls, were now portrayed by light, entering the sanctuary through stained glass. Gothic architecture was the answer to this challenge.

The Gothic style emerged as medieval engineers experimented with the elements of buttressing, vaulting, and arches on smaller buildings, learning from their successes and from their failures as well. In 1144 the great builder Abbot Sugar of Saint-Denis brought these structural elements together as a unified whole. He knocked down the Benedictine abbey's dark basilica and replaced the choir in the Gothic style. Though small by the standards of later cathedrals, this striking new structure impressed visitors from Paris, Chartres, Reims, and Beauvais. They went on to scale up Gothic structures to ever-increasing sizes in their quest for maximum light and maximum height.

Toward this goal, the cathedral builders engineered the replacement of the massive walls that had once supported the roof with thin columns of stone. They succeeded in reducing the pillars' girth and placed them farther apart to allow the expansive windows of brilliantly colored glass to appear where stone once would have stood. With these structural innovations the glass expanses became large enough not only for the portrayal of religious themes; figures in stained class celebrated the civic pride of clerics, kings, merchants—and craftsmen. Even more than space, the requirements for light at well as height were the design criteria that arose from the religious fervor and civic pride of northern France's cities. These were the criteria that drove the engineering innovations of the Gothic structures.

Building a cathedral was a massive undertaking, requiring enormous resources. The civic pride of the growing populations of the free cities along the trade routes of northern France combined with the religious commitment of the age of faith to secure the wherewithal necessary to bring these structures into being. Bishops took an active interest in cathedral construction, many acting as sponsors or patrons. Practical charge and final authority for these great building projects rested not with the bishop, however, but with the cathedral chapter, headed by a dean and made up of canons, who were priests and in some cases dignitaries. Organized to be self-perpetuating, the chapter provided continuity from bishop to bishop over the generations required to complete a cathedral. It resembled the building commission of a modern city. Everything concerning the building and upkeep of the church fell within its responsibility. The chapter raised the money to finance construction; it also appointed the project overseer and adjudicated problems of finance, purchasing, and personnel.

The chapter's canons continually sought new sources of funds. Construction projects of such magnitude required raising a continuous supply of money for materials and wages, for if resources were not forthcoming, work stopped. The later stages of building were most costly, as the shapes of the stones for parts such as vaulting were more complex and thus more expensive to carve. The stone construction of the highest vaulting required exceedingly elaborate lifting devices. At Chartres the main building campaign lasted forty years, but the construction took substantially longer.

The canons raised money from taxes, including taxes on themselves, and from gifts and appeals. Every stratum of society contributed to the construction. The tithes of nobles went toward construction, as did dwelling taxes imposed on the local citizenry. Craftsmen and merchants contributed. Guilds of goldsmiths, masons, carpenters, butchers, bakers, and tanners donated stained-glass windows depicting the pride and joy of the medieval craftsman at work. Outside the town, money was also raised from the rural tenancy; the financing of the Chartres cathedral depended heavily on the prosperous arable land surrounding the city. Contributions also came from farther away, from all over France, and from England—from Richard the Lion-Hearted and the archbishop of Canterbury.

The church frequently authorized those responsible for construction to grant indulgences to contributors of large sums to the building funds. In that age of faith, pilgrims who came to see relics housed at the construction site also contributed, and at times the relics were sent on tour to raise more funds. On occasion the citizenry contributed to the effort by voluntarily dragging carts of stone to the building site. But periodically things did get out of hand. In France, the taxes sometimes became so unbearable that the townspeople rioted. During the thirteenth century such riots caused delays of several seasons in the construction of cathedrals at Reims and Beauvais.

❀   ❀   ❀

For all the architectural acumen needed to achieve the superb aesthetic appeal of the cathedral, building a cathedral required above all the direction of an astute structural engineer of both rare talent and considerable daring. For in designing the cathedral he was defying gravity to an unprecedented degree. The technical challenges facing the medieval architect-engineers in bringing these magnificent structures into being are awe-inspiring.

Stone was the all-encompassing building material of the cathedral builder. Stone has enduring properties, for unlike wood it neither burns nor rots. Its other properties, however, presented severe structural challenges. For stone, unlike wood, has greatly differing strengths in tension (i.e., stretching) and compression. In compression it is nearly uncrushable. A column of stone masonry could be built a mile high if it didn't tip over, and even under this great weight the blocks at the base would not be crushed. But stretching, or tension, is a different story. Under tension stone is substantially weaker than wood. This is most apparent in the behavior of a beam. In the bending of a beam from the weight of a roof, or even under its own weight, the uppermost part is compressed, but the lower part is stretched and thus in tension. As a result, a stone beam of significant length would crack and break, even if it were supporting only its own weight. Thus whereas massive wooden beams can support heavy roofs, the idea of a stone beam is practically an oxymoron. For this reason Greek temples, as pleasing as they may be to look at from afar, contain no significant interior space. Their supporting columns had to be placed very close together; otherwise, the very short lengths of stone that lie atop them, called lintels, would develop too much tension and break.

The architectural engineers of Gothic cathedrals and their predecessors who built Romanesque cathedrals dealt effectively with the limitations of stone. They avoided the use of beams in favor of the arch. If properly proportioned, the arch places all of the stone blocks that compose it in compression as they push against one another under their own weight. The compression within the stone prevents it from cracking, while compression across the rough surfaces between blocks keeps them from sliding out of place. Thus the magnificent Gothic cathedrals are essentially stone block castles, held together only by the geometry of their arched structures. Medieval mortars acted as sealant to retard water seepage and erosion, but they had virtually no structural strength, and unlike wood, stone blocks could not be easily secured by pegs or other fasteners.

Roman engineers, as widely known, used semicircular arches extensively in building bridges and aqueducts. They also built domed buildings; structurally, the dome is a natural extension of the arch, its figure obtained by rotating the arch about its vertical axis. It was, however, the cathedral builders who brought arched configurations to their highest form with the related innovations of the Gothic arch, the ribbed vault, and the flying buttress. These three new structural forms, as we shall see, embodied the aesthetic impact of the Gothic cathedral. But to appreciate that impact we must start with earlier, Romanesque cathedral construction and work our way forward

The earliest churches utilized barrel vaults, such as shown schematically in figure 8, to provide a stone roof. The vault is just an arch extended in the transverse direction. The barrel vault, however, posed a problem: the weight of the arch pushes outward on the walls as indicated in the drawing. Hence, the wall must be very thick to resist this outward thrust, and the taller the walls are, the thicker they must be made. Moreover, since larger windows would weaken the walls, they must be kept to a minimum. Thus larger openings were possible only at the ends of the barrel. Whereas this was acceptable in Mediterranean climates, where the intense sun was the enemy of the interior, in the north the criteria of maximum height and light called for both higher walls and much larger windows.

The first step toward a solution was the cross vault, named after the intersection joining two barrel vaults placed perpendicular to one another, as shown in figure 8. This configuration concentrated the weight and the outward thrust at the structure's four corners. Whereas the corners of these cross vaults, also called "groin vaults," must be buttressed, the walls in both directions are free of the roof's weight, allowing larger openings to be made for windows. Such

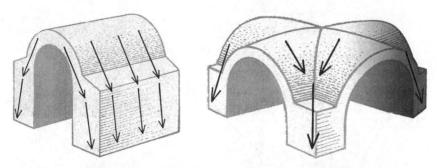

Figure 8. Thrust transmission through barrel (left) and cross (right) vaults.
(Drawing by Irving Geis, in J. H. Acland, "Architectural Vaulting,"
Scientific American 205, no. 5 [1961]: 148. Rights owned by
Howard Hughes Medical Institute. Reproduction by permission only.)

vaults, placed in a line, could then form a sanctuary. This configuration, nonetheless, was still of limited flexibility, for the cathedral floor plan must consist of a series of squares. More importantly, the problem of buttressing the columns against the outward thrust on the barreled roof required a solution. These limitations notwithstanding, arches and barrel vaults allowed the Romanesque cathedrals of the early Middle Ages to begin the reach for height and light.

It was the combination of the Gothic, or pointed, arch with the cross vault that led to the "ribbed vault" that gave the Gothic builders the flexibility they desired. As figure 9 indicates, the ribbed vault results from replacing the semicircular with the pointed arch. This allowed construction of the ribbed vault over a rectangular floor space with more flexible proportions. Since the arch height no longer needed to be one-half of the diameter, the designer could vary floor dimensions by making the proportions of the pointed arches different in the two directions. The builders first erected the crossed ribs and then completed the shell of the vault. Once completed, the shell no longer depended on the ribs for support, and the vaulting ribs became one of the cathedral's purely decorative elements.

Structurally, the ribbed vaulting transmitted the thrust from the roof's weight directly to the pillars, called piers, at the vault's corners. Since the outer walls of the cathedral no longer carried the roof's

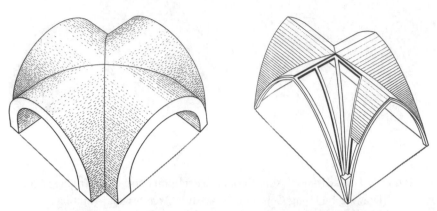

Figure 9. Comparison of cross (left) and Gothic ribbed (right) vaulting.
(Drawing by Geis, in Acland, "Architectural Vaulting," p. 146.
Rights owned by Howard Hughes Medical Institute.
Reproduction by permission only.)

weight, vast expanses of stained glass could largely replace them. But a big challenge remaining was to buttress the piers against the outward thrust transmitted from the roof's weight. Moreover, the farther the piers were placed apart to maximize the window areas, the greater the concentration of outward thrust on each pier. Buttresses, or extensions of the cathedral perpendicular to the walls, would solve the structural problem, but they also would create an aesthetic difficulty. For sufficient strength, the buttresses needed to reach nearly as high as the edge of the roof gable and extend quite far out from the wall. But such massive forms would block the light from entering the windows from all but a small range of angles.

The architect-engineers of the high Gothic responded by cutting away all of the unnecessary stone from the buttresses. They created flying buttresses, such as shown in figure 10, consisting of thin arches geometrically constructed to carry the roof's thrust safely to the ground with a minimum blockage of light. The results were dramatic. At Chartres a double layer of thin, buttressed arches, shown in figure 11, are connected in a lacelike structure that supports the great weight of the cathedral while minimally blocking light. One might say that the designers created them to provide maximum support with minimum stone, in order to resist maximum thrust while minimally blocking light.

❀   ❀   ❀

What do we know of the builders who defined Gothic architecture through these structural innovations, of those who directed the design and construction of the cathedrals? Much less than we would like, for knowledge of their tools and methods has not survived nearly as well as the stone monuments they created. Through the historical records and artifactual evidence, however, some of the characteristics of these medieval architect-engineers can be understood.

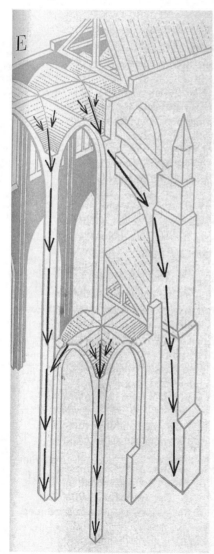

They were not called engineers or architects, but master masons. For in the Middle Ages—unlike today—there was no distinction between designer and builder. The master mason came up through the ranks from apprentice to journeyman to master, learning all of the details, beginning with the cutting and forming of stone. But much more was required to be-

Figure 10. Thrust transmission through a flying buttress. (Drawing by Geis, in Acland, "Architectural Vaulting," p. 149. Rights owned by Howard Hughes Medical Institute. Reproduction by permission only.)

come a cathedral builder. Having gained extensive practical knowledge, a particularly talented master mason would leave the physical labor of cutting and forming stone behind at some stage in his career to enter the tracing house. There he would expand his knowledge of geometry and other subjects and learn to produce the plans from which cathedrals were built.

Located at the building site, the tracing house was where the

Figure 11. Flying buttresses of Chartres cathedral. (From Jeane Gimpel, *The Cathedral Builders* [London: Random House / Pimlico, 1993], p. 145. Reproduced by permission.)

design of the cathedral took shape. There the master mason employed the primary tools of his trade: the compass for constructing circular arcs; the graduated ruler for measuring distances, and the "square" for drawing right angles. With these he created the intricate geometrical patterns for the cathedral and conveyed them to the stonecutters and masons, who would bring the building into existence. Drawings were produced section by section as work proceeded. Some were made on parchment, which was expensive, and so when no longer needed the drawings often were erased and the parchment reused. Not many plans survived, but one such drawing from the cathedral in Cologne that does remain is drawn in ink and fills a sheet of parchment ten feet long and three feet wide.

The master created floor plans, sections, and details. With compass and rule, he set out each working drawing precisely to scale. Not a single unnecessary line appeared. The mason used centerlines and other abstractions to reduce the effort required. He would lay out the full sized drawing for each block of stone masonry on the tracing house floor. Assistants would create templates by tracing the master's outlines onto wood and cutting it to shape, and the templates in turn guided quarry masons in cutting the stone.

Before graduating from the tracing house to take full responsibility for constructing a cathedral, the master mason had to demonstrate a high level of organizational skill. This was essential to assure that he could properly implement the design in stone. At any one time he might direct the work of several hundred artisans. Supervision was complicated because some masons worked at the construction site while others were located at quarries. Then, as now, the sequencing of construction work was very important, and slow communications and the great difficulty in transporting the numerous stone blocks further complicated these tasks.

The master was responsible for estimating the quantities of material and amount of labor required to accomplish all tasks involved in construction. For some projects he directed the financial administration as well. He had to find quarries where suitable stone was available, stone that both worked and weathered well, stone that was soft to quarry but that hardened with age. He also oversaw the great challenges in transporting the large masses of stone from quarry to construction site. Although a stonemason by background, the master had to know carpentry as well, not only to supervise the slate or lead-covered wooden roof that rose over the vaulting but also to design the ever-changing scaffolding necessary during construction. What is more, he needed expertise in the varied mechanical devices for lifting, sawing, and other operations. In some cases success required him to engineer boats and docking facilities in order to transport stone from the quarry to the cathedral site.

The cathedral builder worked closely with the canons and the chapter dean. They paid him well and provided him housing and other amenities. Before rising to this status he would have had to become literate as well as traveled, despite the limitations of medieval transportation. The more renowned cathedral builders were sought after and would be called to other construction sites as engineering consultants when particularly perplexing decisions were under consideration. The chapter frequently found it necessary to bind the master mason in charge of the project with contracts that stipulated the time that he would stay at that site. Otherwise he might take on too many projects and be absent too frequently to provide adequate guidance.

The cathedral builders' work expressed more than the religious fervor of medieval times. Intense civic and regional rivalries also challenged them to engineer structures that reached higher and higher toward the heavens. In 1163 the vaulted ceiling of Notre Dame in Paris was constructed to a height of 108 feet. But then in 1194 the builders of Chartres set a new record with the vault of 120 feet. They in turn lost the record to the cathedral at Reims in 1212 with its 125-foot vault, and then to the cathedral at Amiens in 1221 with a vault height of 139 feet. In 1225 the Beauvais cathedral's vault set an all-time record of 157 feet. This medieval competition in structural engineering has often been compared to that between Chicago and New York in the building of twentieth-century skyscrapers.

The cathedral builders ventured into the unknown. More height and more light required more elaborate structures with proportions and geometries leading to thinner, taller, more graceful structural members. How did they maintain structural stability and avert collapse? In *The Stone Skeleton, Experiments in Gothic Structure,* and other texts, present-day engineers have proposed structural theories and the scientific explanations of how these medieval treasures seemingly defy the forces of gravity.[2] The medieval architect-engineers, as noted earlier, designed the cathedrals centuries before modern scientific methods began to emerge, long before Galileo published the first text on the strength of materials. How they chose the shapes and proportions of the piers, spires, vaults, and buttresses to avoid collapse while venturing yet higher remains to a great extent a mystery. But some tantalizing clues do exist.

Of the surviving documents perhaps the most insightful is the notebook of Villard de Honnecourt, an architect-engineer who was among the cathedral builders practicing in the thirteenth century. Thirty-three pages, in a well-worn leather cover, are all that remain of a substantially longer work. The book contains drawings on both sides of the pages and notes jotted down on a wide range of subjects. It displays plans, sections, and elevations of buildings; details of masonry and carpentry, construction, and geometrical solutions to problems, machines for lifting heavy weights; and techniques for sawing timber. Labor-saving devices also fascinated him. The notebook also portrays monumental depictions of prophets and other biblical figures that he may have intended to serve as models for cathedral sculptors.

We know that such manuals and pattern books were valuable legacies passed from one generation to the next in the tradition of the medieval stonemasons' guilds. Villard's notebook is instructive as a repository for ideas and engineering techniques. Totally absent are any numerical calculations. The attempts to show the working of mechanical devices are not drawn in perspective, a technique unknown in medieval Europe, but rather in a flat, two-dimensional form that appears strange to the modern eye. The drawing of a sawmill operation, reproduced in figure 12, is typical. The absence of perspective drawings notwithstanding, the notebook contains

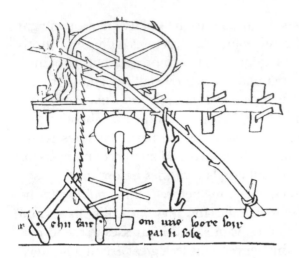

Figure 12. Villard de Honnecourt's water-powered saw. (From R. Willis, ed., *Sketchbook of Villard de Honnecourt* [London, 1859].)

many rules of geometrical proportion as well as intricate geometrical representations of floor plans, elevations, and other constructs. This is key, since geometric proportion, not calculations of material strengths or deflections, was central to the design of stone masonry structures that were stable and safe. The compass and square were the essential tools with which the architect-engineer applied his deep geometrical knowledge to create a cathedral.

The properties of stone and geometry-based design go together. Stone is so strong in compression that the danger of crushing can be ignored, and both stone and mortar are so weak when stretched in tension that we may assume that they have no strength at all. The primary challenge in cathedral design was not that of strength but of assuring stability, of making sure that that no part of the building would tip and fall. Consider the analogy of a child's block tower. There is no chance that the blocks will be crushed. Thus in building a tower, the primary challenge is in placing the blocks on top of each other in a stable arrangement so that the tower does not tip over.

In a much-compounded form, the ribbed vaulting and flying buttress presented a similar challenge to the cathedral builders as they oversaw the placing of thousands upon thousands of carefully chiseled blocks atop one another to form their masterpieces of stone. Scale models would have been helpful, since proportions and not size determine stability, and with a model the stability of a new configuration might be studied without risking lives or financial resources. No evidence survives, however, of structural models built before early in the Renaissance. Moreover, it appears the earliest models were used not as an aid in structural design but for communicating proposed designs to bishops, chapter members, or other patrons.

The theory of "thrust lines," which originated hundreds of years later, shows the flow of the compressive force of the structure's weight of as it is carried toward the earth. The mathematics that allows these lines to be constructed enables an engineer to predict the onset of structural instabilities. Without this knowledge, the medieval architect-engineer relied on design rules gained from trial and error and encoded in the elaborate constructs of geometric proportions that he knew to be safe from training and practice. But the information was in a form that only master masons after years of guild experience were equipped to understand. The books of the craftsmen, passed from generation to generation, contained the geometrical constructs and rules of proportion that were known from experience to result in stable structures. Unlike today's safety factors of strength, which are expressed numerically, the medieval safety factors were geometrical, encoded in the proportions of the drawing of the engineers' notebooks.

Even with caution, and all the wisdom that their experience could command, the builders' efforts sometimes met with failure. The point of greatest danger occurred when a vault or a buttress was completed and the wooden bracing that supported it during construction was removed. There always lurked the possibility that the structure would come tumbling down. Even if a newly completed section of a structure stood, dangers remained that might cause collapse at a later date. Windstorms created large horizontal forces on a cathedral's exposed faces, and they would tend to tip the cathedral away from the wind if enough geometric safety margin was not present to resist these forces.

More insidious was the settling of foundations, for while stone may be very strong in compression, the earth is less so. And a slight unevenness in settling could cause structural instability. Thus the builders inspected their work as the cathedrals went up, looking for the formation of cracks in the mortar or stone that might foretell eventual collapse and making design corrections midstream as they proceeded.

At times, however, the medieval builders' reach exceeded their grasp, and collapse occurred. Likely the most spectacular of these took place at Beauvais. The most daring of the Gothic architecture to adorn northern France, the cathedral's high vaults stood 158 feet above the ground, the highest yet to be reached. The intricate structure, supported by an elaborate pattern of flying buttresses, with each thrust carefully balanced by a counterthrust, approached most closely the ideal of a stained glass building supported only on a thin skeleton of stone. But in 1284 the vaulting of the unfinished cathedral came crashing down.

The collapse was not the result of an earthquake or windstorm upsetting the structure's delicate stability; it appears, more likely, that over twelve years uneven settling of the building's foundation caused an imbalance to develop, causing the vaulting to collapse. It required fifty years for repairs and modifications to be made and for rebuilding to take place, but by 1337 the cathedral was restored to its precollapse state of completion. The Hundred Years' War caused a long hiatus, and not until 1500 did construction resume. The cathedral's great tower was completed in 1569, rising nearly five hundred feet. It was a daring accomplishment. But the tower's thousands of tons of stone rested on four columns and placed unprecedented pressure on its foundation. And, indeed, two years following completion, the tower began to lean. The chapter called master masons from Paris for consultation. But before corrective action could be taken, the tower fell. Large loss of life was narrowly averted, as an Assignation Day procession of clergy and parishioners had just exited the cathedral. Repairs were made, but the tower was never rebuilt, the cathedral never completed.

The setbacks of Beauvais notwithstanding, the height and light of Gothic architecture spread from its cradle in northern France across Europe to England, Spain, Germany, and Austria. The architect-engineers learned from failures as well as successes and continued to innovate, creating cathedrals tailored to the varied cultures and climates in which they worked: witness the steep roofs and towering spires at Cologne, Ulm, and Vienna and the elaborate vaulting of Exeter, Gloucester, and Cambridge.

❀   ❀   ❀

Standing before Our Lady of Chartres I found myself comparing the cathedral's builders with their counterparts responsible for erecting the skyscrapers of the twentieth century. Their profession differed greatly from that of today's engineers and architects. With tradition deeply rooted in the medieval craft guilds, they accomplished their prodigious tasks without the tools considered so central to engineering today. Natural philosophy, the science of medieval times, focused on astronomical observation and philosophical argument and had little relevance to the building trades. Methods of investigation that integrated experimentation with mathematical analysis to provide useful insights and data would not come until four centuries later, in the wake of the scientific revolution. The builders lacked the means to visualize and communicate three-dimensional constructs through drawings in true perspective, for that form of expression was unknown to them. The available mathematics was confined to geometry. The Arabic system of numbers had yet to penetrate Europe, and with only Roman numerals at their disposal, attempts to perform even simple multiplication or division would have been an exceedingly difficult undertaking. The formulas and equations so ubiquitous in engineering today lay even farther in the future.

Though lacking most tools of modern practice, the medieval architect-engineers' profession encompassed far more than the trades of skillful artisans who produced wagons, wheels, and the profusion of other artifacts of everyday life. For whereas the artisan crafted each new wagon or wheel to closely replicate the last, the cathedral builders ventured into the unknown. The structures they erected were unique, each requiring innovation to achieve height and light beyond what had existed before. Each presented intertwined challenges in structural design, construction method, and project management. Each presented problems whose solutions demanded talents and training that far exceeded what the artisan's work required.

At the time these great cathedrals were built the distinction between architect and engineer did not yet exist. That would develop later, during the Renaissance, when architecture became the profession focused on the artistry of determining a structure's style, shape, and symbolism, while engineering concentrated on building it and making sure that it would stand. But the medieval masters did both. They conceptualized cathedrals aesthetically as well as structurally and communicated their designs through drawings to the artisans who would build them and to the patrons who sponsored them. They estimated the labor and materials needed to erect the cathedrals and negotiated with the chapter to obtain the required resources. They scheduled and coordinated the activities of vast numbers of craftsmen from numerous trades and organized transport of materials from quarry and forest to building site.

Above all, the cathedral builders had to be adept at solving problems encountered in design and construction. Through experience, drawings, and trial-and-error methods, they strove to meet the demands for greater height and increased light while limiting risks and husbanding resources. Already their habits of mind and processes of design formed the foundation that would later coalesce through the advances of the Renaissance, the scientific revolution, and the industrial revolution to create the engineering of today.

Four

Patrons of Progress

Until quite recently in human history much technological advancement had derived from the accumulation of nearly imperceptible changes made by craft workers. In a sense similar to Darwinian natural selection, artisans replicated the more useful of these changes but discarded others that were less successful. Yet even long before science became an indispensable part of the engineering endeavor, circumstances did arise that accelerated the pace of technological advancement. These resulted in innovations that brought profound changes to the societies in which they occurred. But prerequisite to such technological advances were the resources necessary to finance the risks of engineering innovation, for without that financial backing, even the most modern tools of engineering would have been of limited value. Conversely, provided that they had the time to think, to experiment, to fail, and to try again, the engineers of earlier times accomplished wondrous things, even without the professional tools and knowledge that we consider so essential today.

Only cultures that had risen sufficiently above subsistence level could contemplate the diversion of valuable resources—food, skilled manpower, and materials—to the inherently risky venture of technological innovation. This would tie up resources for extended periods of time, and a useful outcome could not be guaranteed. Innovation was a possibility in a society whose values and power structure determined that marshaling accumulated resources for engineering purposes was a worthwhile enterprise. The drive toward technological innovation required nurturing in a social context in order to provide sufficient manpower, materials, and patience that would allow engineering exuberance to work its wonders.

Thoughts of contrasting civilizations and the values they held concerning technology came to mind when I recently returned to Stuttgart in the Neckar River valley of southwestern Germany. Home to Daimler-Benz, Bosch, Porsche, and other corporate powerhouses, as well as a university with a world-class engineering school and several preeminent research institutes, Stuttgart is a beehive of modern engineering financed through industrial firms, government agencies, and private philanthropy. But looking beyond Stuttgart to the hills on either side of the Neckar reveals outposts of two earlier civilizations and remnants of the technological innovations that they valued and supported. To the east, along the ridges of the Swabian Jura, extend the ruins of the *limes*, the wall and its fortifications that delineated the northern frontier of the Roman Empire. To the west, next to the road leading into the mist-shrouded mountains of the Black Forest, stands the remains of the venerable medieval monastery at Hirsau. Both symbolize the technologies of long ago, developed under very different circumstances and influenced by very different patrons and values.

The *limes* represents a great feat of state-sponsored engineering, which reflected the power of Rome. Stretching across southwest Germany to protect the empire from Germanic tribes, it reappears as Hadrian's Wall across the north of England, where it served to stave off invasions by the fierce Scottish highlanders. The ruins of these frontier walls provide but one example of the Roman Empire's high degree of engineering acumen. The walls were served by an immense system of highways emanating from Rome that helped to maintain control over the vast lands of the empire and defend its frontiers from barbarian tribes. Like the Colosseum, other state buildings, and the aqueduct systems for supplying water to the cities of the more arid Mediterranean regions, these feats of civil and military engineering reflect the empire's achievements. They display its ability to combine vast material and human resources with the organizational skill required to raise technology beyond humble crafts, and that takes it into the realm of systems engineering.

The buildings constructed along the *limes* exhibit more than military fortifications. They, in essence, transplanted the culture of Rome to this remote northern frontier. The ruins of the baths at Schwäbisch Gmünd capture the penchant of Roman citizens of stature for cleanliness and bathing. Roman baths were not equaled in their technological sophistication for well over a millennium. Typically a bathing complex consisted of an undressing room, a cold plunge bath (*frigidarium*), a bath of tepid heat (*tepidarium*), a hot bath (*caldarium*), and an exercise court. Most sophisticated was a room of intense dry heat (*laconicum*) very similar to the sauna of today. The complex structures and plumbing demanded sophisticated hydraulic engineering. Circulation of prescribed quantities of water in each of the baths required careful planning. Execution of those plans in turn necessitated a detailed knowledge of volumes and rates of flow and the ability to control water levels in all of the buildings with great accuracy.

The baths, the reservoirs from which they were fed, and the interconnecting piping presented many challenges; losses from leaks had to be minimized, and polluted groundwater prevented from entering the system. Success came by lining the carefully constructed stone structures with sheets of lead. Lead was also the material of choice for the plumbing. Connecting pipes were formed by hammering strips of lead around wooden poles until the edges made contact and then pouring molten lead along the seam to create a properly fused joint. Water supplied from a spring or stream frequently brought sand or dirt with it, and this was removed to prevent clogging of the plumbing and to meet the purity standards of Roman bath water. To accomplish this, the engineers designed the reservoir to act as a settling tank and arranged the gates, drains, and sluices to provide a system that would periodically wash away the unwanted sediment that otherwise would slowly fill the reservoir.

Perhaps the most innovative engineering, however, was driven by the desire for temperature control and effective forms of central heating. In addition to controlling the temperatures of the water in hot, cold, and tepid baths, the Roman engineers provided ambient heat for bath, dressing, and exercise rooms to meet Roman citizens' expectations of comfort, even in the cold of winter. The heat came from charcoal-fueled furnaces placed under large copper boilers. These provided a constant supply of hot water. For room heating the engineers developed the hypocaust, a natural circulation system illustrated in figure 13. It drew hot air from furnaces to a cellar cavity located under the occupied room. After the air heated the floor, natural buoyancy drew it upward though vertical flues or chimneys in the walls. The air heated the walls as it rose higher before exhausting through openings in the roof. Once a hypocaust room reached the desired temperature, the thick masonry of the walls, floors, and ceiling ensured that the heat would not escape easily.

These bathing facilities indicate the level of comfort and recreation enjoyed by the empire's citizens. But the term *citizen* must be employed very carefully, for in the civilizations of antiquity, the citizenry accounted for only a small fraction of the population. The vast mass of humanity consisted of slaves and indentured servants. The muscle power of slaves built the massive stone edifices, aque-

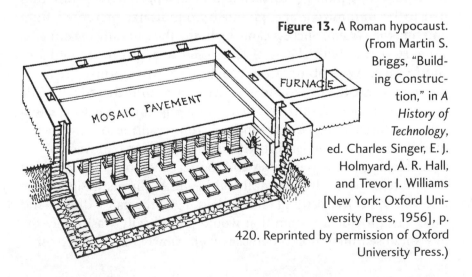

Figure 13. A Roman hypocaust. (From Martin S. Briggs, "Building Construction," in *A History of Technology*, ed. Charles Singer, E. J. Holmyard, A. R. Hall, and Trevor I. Williams [New York: Oxford University Press, 1956], p. 420. Reprinted by permission of Oxford University Press.)

ducts, and highways. They provided not only the manual labor for building the baths but also prepared the charcoal, tended the fires, and maintained the plumbing of these elaborate structures.

In contrast to the ingenuity and labor expended for military power and for the grandeur of the state—and for the comfort of the restricted class of citizen—the Romans devoted exceedingly few resources to developing technologies that would lessen the toil of the slaves. Why would they? Slavery was widespread, thanks in great part to armies and peoples conquered by the Romans. Since the society's drudgery was borne by those with no political power or financial resources, little attention was given to their plight. The technologies pursued reflected this societal structure and the view of the master toward servant and slave. The unfortunate masses provided such a plentiful supply of cheap and flexible labor that the privileged classes saw no advantage and felt little need for the development of labor-saving technologies.

Even the labor-saving technologies that were known to the Romans tended to languish, for they had little incentive to employ them widely and practically none toward pursuing their improvement. Animal harnesses remained primitive, choking the horses, and greatly limited the substitution of animal for human muscle. The Romans understood various clever devices and automata, but only in later ages would they find application for labor-saving technologies. In Rome they served only as toys and gadgets for the amusement of the privileged few. The elite paid little attention to devising more effective sails that would substitute wind for the muscle of the galleon crews; they understood the waterwheel's labor-saving potential but rarely put it to use.

The waterwheel's history is indicative of this tendency. The first-century BCE writings of Roman architect-engineer Vitruvius provide a description of the undershot vertical waterwheel and the associated gearing needed to change the direction of rotation so that it could grind grain. This technology required a substantial level of engineering sophistication. But not until five hundred years later, when the Roman Empire was contracting and under siege, when the supply of slaves from conquest was rapidly vanishing and labor shortages were growing acute, did the Romans begin to seriously apply waterwheels to the grinding of grain and other labor-saving tasks. The waterwheel complex near Arles, in the south of France, is the most significant example of the increased attention given to waterpower toward the end of the Roman era. As the empire's power declined, the buildings, baths, walls, aqueducts, and highway systems could no longer be maintained by conscripted labor. And with the collapse of the western empire, the infrastructure for state-sponsored technological development fell into ruin.

On the other side of the Neckar, the Black Forest monastery at Hirsau was an outpost of the medieval civilization that developed out of the feudal system of the Dark Ages. It symbolizes technological accomplishments that differ as much from the Roman baths along the *limes* as the monks' monastic lives differed from those of the Roman citizenry. Tumultuous changes took place between the time that the Roman baths fell into disuse and the monastery was established at Hirsau. Population numbers in western and central Europe underwent rapid decline in the sixth and seventh centuries CE, at the onset of the Dark Ages. The centralized power of the imperial state had dissipated, and small enclaves of feudal society took its place. Equally as important, classical thought, which viewed the crafts and manual labor with disdain, was supplanted by Christian doctrine that eschewed slavery and saw crafts and labor in a new light.

Early in the sixth century Saint Benedict established an extensive set of rules for monastic brotherhoods that was widely accepted throughout western Europe. Spreading northward, these rules codified two central elements in the Western monastic tradition: the first was the drive for self-sufficiency, thereby protecting the monastery from worldly corruption; the second established the dignity of manual labor and asserted that such labor was a form of worship. The desire for self-sufficiency and the belief in the dignity of labor came together in regulating the life of Benedictine monasteries. For this end, Saint Benedict recommended architectural layouts in which all necessary activity, from grinding grain to gardening, could be located within the confines of the monastery. Benedictine rules divided the monks' days into times for prayer and meditation, for reading and study, and for manual labor. Since servitude was contrary to Christian beliefs, the monasteries faced a dilemma in requiring the monks to sustain themselves with their own labor while reserving adequate time for study and meditation.

Here then was a great incentive for the development of labor-saving technologies. Provided with the pool of skilled, motivated, and disciplined labor found in the monasteries, the abbots were able to husband their resources to pursue the development of labor-saving devices. With time and materials to experiment, to fail, to learn from those failures, and to try again, the monks began their quest for technologies that would reduce physical drudgery and free time for meditation and scholarship. In that quest, the Benedictine monks became the power engineers of the early medieval era. They engineered the waterwheel to power a range of labor-saving devices on a scale unheard of in classical antiquity.

The monks began by improving the Roman undershot wheel described by Vitruvius and adapted it to monastic use. Undershot wheels, such as the one illustrated in figure 14, placed in fast-moving streams developed power from the direct impact of the water. With time, the monks also engineered effective overshot wheels such as the one shown in figure 15. These were most effective where a larger drop in stream level allowed them to effectively utilize the weight of the falling water. Construction of waterwheels and their associated gearing required the monks to become masters in woodworking and mechanical design. They also devised increasingly sophisticated structures, particularly for overshot wheels: races to channel the flow of streams to the wheel and reservoirs to store the water and even the flow over periods of rain and drought. The building of dams, channels, and similar structures to maximize the effectiveness of more powerful overshot wheels required the monks to become accomplished in masonry construction and structural design.

The first widespread use of waterpower was for the nearly universal, labor-intensive task of grinding grain into flour. This task,

Figure 14. An undershot waterwheel powering a grindstone.
(From James Burke, *Connections* [Boston: Little, Brown, 1978], p. 86.
By permission of Little, Brown and Company.)

perhaps more than any other, accounted for hours of drudgery in the societies of antiquity. Mechanized grain grinding required translating the waterwheel's rotational motion from horizontal to vertical and speeding up the slow rotation of the waterwheel to spin the grindstone at a much higher rate. The monks perfected gearing, such as is illustrated in figure 14 above, to accomplish this task. While engineering the techniques for milling grain effectively, they also sought other ways in which waterpower could replace human muscle. A number of uses necessitated converting the rotational motion of a waterwheel to reciprocating or hammering motions. To this end they engineered systems that effectively utilized the cam: figure 15 shows two assemblies of cams attached to the shaft of a waterwheel, one driving a trip hammer and the other a bellows.

Waterpower also led to the mechanization of the fulling process. An important step in the production of cloth, fulling was pervasive in preindustrial society. Following weaving, fullers cleansed,

Figure 15. An overshot waterwheel powering a hammer and bellows. (From Burke, *Connections*, p. 87. By permission of Little, Brown and Company.)

scoured, and thickened cloth by beating it in water. But where men's muscle once pummeled the cloth, now wooden hammers, lifted and dropped by means of a revolving cam attached to the spindle of the waterwheel, performed the task. Water-powered hammers also revolutionized the forging of iron, for their power greatly exceeded the blacksmith's physical strength. With the engineering of increasingly elaborate gearing, cams, bellows, pumps, and other ancillary mechanisms, waterwheels also replaced human muscle in crushing olives and grapes, tanning leather, pumping water, smelting iron, making paper, and performing other tasks.

The power and versatility of the waterwheel was so impressive that its use spread throughout the monasteries of Europe. Historical records provide some specifics. In 855 the abbey in the Pas-de-Calais had three water mills. In 845 the monastery of Montier-en-Der operated eleven water mills on the river Voire, while the monastery of Saint-Germain-des-Prés owned fifty-nine. Even at Silvanes, a relatively small abbey, monks built two dams and used four water-wheels for milling and fulling. From the ten wheels owned by the Welsh abbey of Whitland to the sixteen water mills operated at the monastery of Grandselve in the southwest of France, waterwheels found use in geographically dispersed locations. Some projects entailed massive excavation. Near Paris at Royaumont, monks constructed a tunnel more than one hundred feet long to supply water to wheels that ground grain, tanned leather, fulled cloth, and worked iron.

Of the Benedictines, the Cistercian order played a particularly important role in expanding the use of waterpower to the remote areas of Europe. Founded by Saint Bernard early in the twelfth century, the order grew to over five hundred monasteries over the next hundred years. In monasteries from Sweden, Scotland, and Wales to Portugal and Hungary, the monks relied on the power of flowing water to meet their material needs. Abbot Arnold of Bonneval's poetic tribute to the value of water summarized its many uses at the rebuilt Cistercian monastery at Clairvaux, France:

> Entering the Abbey under the boundary wall, which like a janitor allows it to pass, the stream hurls itself impetuously at the mill where in a welter of movement it strains itself, first to crush the wheat beneath the weight of the millstones, then to shake the fine sieve which separates flour from bran.
>
> Already it has reached the next building; it replenishes the vats . . . to prepare beer for the monks. . . .
>
> The stream does not yet consider itself discharged. The fullers established near the mill beckon to it. In the mill it had been occupied in preparing food to the brethren; it is therefore only right that it should now look to their clothing. It never shrinks back or refuses to do anything that is asked of it. One by one it lifts and drops the heavy pestles, the fullers' great wooden hammers. When it has spun the shaft as fast as any wheel can move, it disappears in a foaming frenzy. . . .
>
> Leaving here it enters the tannery, where in preparing the leather for the shoes of the monks it exercises as much exertion and diligence, then it dissolves in a host of streamlets and proceeds

along its appointed course to the duties laid down for it, looking out all the time for affairs requiring its attention, whatever they might be, such as cooking, sieving, turning, grinding, watering or washing, never refusing its assistance in any task. At last, in case it receives any reward for work which it has not done, it carries away the waste and leaves everywhere spotless.[1]

Medieval monks were the power engineers to spearhead the development of waterwheels and thus relieved the burden on human muscle for many tasks of monastic life. Not surpassingly, the benefits of the waterwheel spread with time from the monasteries throughout the medieval world. The peripatetic existence of artisans and craftsmen brought them into contact with these technological innovations. As lay craftsmen came into increased contact with the monks, they assimilated the new methods and brought them outside the monasteries. Feudal aristocrats adopted the use of waterwheels, for with waterpower they could better face their own challenges of self-sufficiency. They could more easily provide large numbers of serfs with the necessities of life while maintaining enough surplus to pursue their chivalrous activities and subsidize the mounted knights of that age.

In 1086 William the Conqueror sent tax commissioners to make a survey and collect information on water mills in thirty four English counties. Their *Domesday Book* indicates the extent to which waterpower had already penetrated medieval life. They recorded more than five thousand mills in use. On some rivers there were as many as thirty water mills in a ten-mile stretch. With the passage of time, the revival of trade and urban life gave rise to an influential merchant class concentrated in urban centers. Waterwheels contributed substantially to the economic growth and independence of the free mercantile cities that became the cultural centers of the high Middle Ages.

Waterwheel use in medieval Europe became so widespread that disputes often arose over mooring and damming privileges. Mill operators frequently placed undershot wheels on boats and moored them between the piles of bridges where the flow was most rapid and yielded the highest rates of rotation. But increasing numbers of such boats brought congestion and interfered with river navigation. Dammed streams increased the productivity of waterwheels, for they created the higher drops needed for large overshot wheels and provided a reservoir to even the seasonal variations in flow. However, owners frequently built the dams too close together, and the reservoir from a downstream dam would interfere with the drop available to its upstream neighbor. Eventually more centralized engineering efforts overcame such conflicts by creating larger hydropower complexes. The mazes of floating undershot wheels, anchored to bridge abutments, gave way to the building of large dams to drive complexes of more powerful overshot wheels. Instead of locating floating wheels in harmony with natural flow patterns, authorities channeled, dammed, and otherwise diverted rivers to facilitate maximum power production.

The engineering of waterwheels and of the associated cams, trip hammers, bellows, and other mechanical devices fostered the development other power sources in areas where flowing water was inadequate. In low-lying coastal areas, rivers were too sluggish for waterwheels to flourish. In the twelfth century, however, tidal mills appeared to fill the void. A dam with gates was placed across the mouth of a bay or river. The dam captured the water of the incoming tide as the rising level of the ocean pushed open the gates. The pressure of the impounded water then closed the gates as the tide began to recede and thus created a reservoir. During low tide the impounded water would pass over waterwheels, producing power as it dropped to the level of the sea.

Windmills found far more extensive use than tidal mills. First appearing in the twelfth century, they spread across Europe, becoming major power sources, particularly in lowlands where streams were too sluggish for the effective use of waterwheels and in the north, where windmills could operate year-round, even while rivers were clogged with ice. The Western windmill, whose blades rotate along the vertical plane, was distinct from the earlier horizontal mills of the Middle East. Originating from waterwheel engineering, the most innovative feature of the windmill was in the blades, which were designed to be turned by air instead of water. Waterwheel know-how provided a basis for much of the gearing and the elaborate mechanisms developed to rotate and control the mill with changes in wind speed and direction.

The medieval engineers created a society that for the first time in history succeeded in substituting the power of machines for human drudgery, a society whose accomplishments no longer rested on the backs of toiling slaves. They harnessed water and wind to enrich the population and to allow it to expend its energy on other pursuits. Where did the resources come from to drive these technological innovations? The answer is complex, multifaceted, and still not fully understood. But it does seem certain that Benedictine monks can stake the claim to being among the West's earliest power engineers. They persuaded their abbots to use the monastery's resources to sponsor the work that initiated a medieval power revolution.

The vast resources of the empire financed Roman engineering, and the communal resources of monasteries sponsored the early development of waterpower in medieval times. Examination of technological history reveals numerous other times and other places where engineering innovation had such enormous impact that it materially changed the society in which it arose. Certainly such momentous changes were not likely to result from the nearly imperceptible pace of technological evolution resulting from subsistence craftsmen replicating artifacts. Thus each marked technological innovation occurring over a measurable period of time raises the question of who provided the resources needed for the engineering development efforts? Who backed the project financially until it could be brought to fruition and its rewards gathered? Whether the sponsor was a royal family, a capitalist entrepreneur, a corporation, or a government, each situation was unique. To provide some perspective on the changing relationships between innovators and their sponsors, we briefly examine the lives of four engineers, each of whose work was instrumental in bringing revolutionary technology into existence. We consider first Henry the Navigator, then James Watt, and finally Wilbur and Orville Wright.

Almost exactly a century before the famous voyage of Christopher Columbus, the prince who became known as Henry the Navigator was born in 1394, the third son of King John of Portugal. Henry eschewed the effete life of the court and chose to lead an ascetic existence. He employed his considerable personal fortune in land holdings, fishing monopolies, and other royal privileges to finance the development of the technological system for oceanic expeditions that was to change the world.

The site where this system was engineered lies to the west of Gibraltar at the very southwestern tip of Europe. There, where the Sagres Peninsula juts out into the Atlantic, Prince Henry created a remarkable community of shipbuilders, cartographers, instrument makers, scholars, and sea captains. The Sagres community looked away from Europe and the known waters of the Mediterranean to the challenges of the far reaches of the Atlantic and the legend-shrouded seas of the uncharted west African coast. For forty years following its establishment in 1419, Sagres headquartered the systematic planning of voyages and preparation of expeditions to ever-farther reaches of the unknown oceans. It, along with the neighboring port of Lagos, was the West's predominant center for innovation in cartography, navigation, and marine engineering.

The Sagres community systematically tackled the three obstacles to exploring the worlds beyond the Atlantic horizon. The cartography of the fourteenth century was shrouded in theology and metaphysics, complicating the collection and interpretation of data from which modern mapmaking would evolve. The maps' beautiful symmetry, in harmony with the symbolism of metaphysical beliefs and biblical teachings, served well their makers' theological intents, but they were of little value in constructing realistic charts of proven discovery and reproducible knowledge. The cartographers of Sagres, however, freed navigational cartography from these impediments to create objective maps from sea captains' logs, firsthand observations, and the careful plotting of coastline contours. They banished the age-old practice of rationalizing legend by including mythological locations on maps. Only by objectively synthesizing what was known could they plan voyages into the uncharted waters of the west African coast and beyond.

At Sagres, Prince Henry also utilized the collective experience and scholarship of a seafaring nation to systematize more rational means for navigating uncharted waters. The compass was known but mired in superstitious distrust and credited by seafarers with mysterious powers. Prince Henry dispelled the mysticism from the compass and instituted its use as an instrument of navigation. The Sagres community developed a simplified instrument for measuring latitude from the angle of the sun; Henry furthermore encouraged the study of the night sky, not for astrology, which was accepted among his contemporaries, but as an objective means of determining time and direction.

Technologically, however, Prince Henry's most memorable accomplishment was the engineering of a ship explicitly designed for exploration. Expeditions on the open ocean required ships very different from those that plied the Mediterranean and the coasts of Europe in the fifteenth century. The most modern ships available were square-riggers called barcas. Their large size maximized the owners' profit in carrying materials or passengers between known destinations. Ships of exploration, however, demanded totally different design criteria. The primary cargo of expeditions into the unknown was the information brought back, for trade and heavy cargoes would not come until later. Thus very large ships were unnecessary, since a ship of exploration needed only to carry its crew and provisions for a long voyage. Size was easily sacrificed. Speed, maneuverability, and ease of repair were far more important. Speed was essential to shorten the burdens of long ocean voyages, and maneuverability was necessary to ply unknown coasts, islands, and straits when sailing far from known ports. Self-sufficiency was essential, and repairs had to be simple enough for the crew to perform.

Most important of all, the design had to maximize the ship's ability to return safely to its home port. For a trading ship to be lost following delivery of its cargo was tragic, but at least a half of its mission had been completed. With the loss of a ship of exploration on its return voyage, however, all was lost. All of the information that it had set out to collect would go down with the ship, and nothing would be added to the store of knowledge needed to plan future voyages. Assuring a safe return voyage was particularly difficult for the explorations the Portuguese planned along the west African coast. The persistent northwesterly winds meant that the outward-bound trip was windward, but the return voyage was spent entirely sailing into the wind. The square-rigged barca, commonly used for carrying cargo, could sail no closer to the wind than an angle of sixty-seven degrees. Thus progress on the return trip would be very slow and harrowing as provisions ran low. More than its other shortcomings, poor performance when sailing into the wind ruled out the barca, if the Portuguese ships were to venture far down the African coast.

Prince Henry's shipbuilders met these challenges by designing a vessel specifically for exploration. It differed most importantly from Portugal's earlier oceangoing ships in carrying the slanted, triangular lateen sail. Arab ships called caravos, which carried cargo off the Tunisian coast, had such sails. In north Portugal on the Douro River, a much smaller and highly maneuverable boat called the caravel employed similar sails. The lateen sail had a great advantage over the square-riggers of the time: it could sail at an angle of only fifty-five degrees to the wind, compared to the square-riggers' sixty-seven degrees. As indicated in figure 16 the distance traveled in sailing into the wind decreased by about one-third, with comparable savings in the time at sea. The lateen sail eliminated several weeks from the trips being planned, with a corresponding decrease in the quantity of provisions required and an increase in the crews' morale.

Henry's shipbuilders engineered a composite vessel, with the maneuverability of the Portuguese caravel and some of the freight-carrying capacity of the Tunisian caravo. They adopted the lateen sails to propel the highly maneuverable hull, which also incorporated the axled rudder from the shipbuilding tradition of northern Europe. The vessel was only seventy feet in length but roomy enough to carry a crew of twenty in addition to their provisions on extended voyages of exploration. Moreover, the craft could sail in shallower water than the cargo ships of the time, allowing the crew to beach it

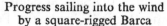

Progress sailing into the wind
by a square-rigged Barca

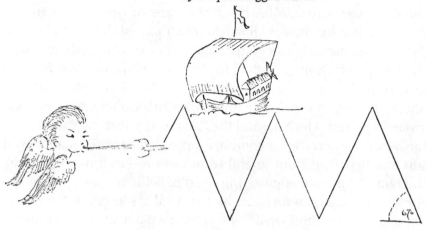

by a lateen-rigged Caravel

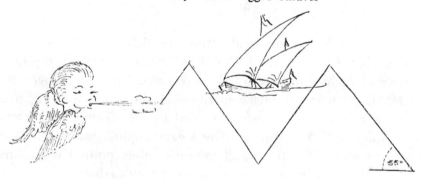

Figure 16. Sailing into the wind: a comparison of square-rigged and triangular lateen sails. (Diagram by Rosemary Greenfield, from John Ure, *Prince Henry the Navigator* [London: Constable, 1977], p. 101. Reproduced by permission of Constable & Robinson.)

with a minimum of difficulty, in order to carry out land exploration or for repairs. This compact and agile craft, the caravel, was a true technological innovation; it became the mainstay of the systematic Portuguese exploration of the west African coast and beyond.

The royal resources invested at Sagres paid off handsomely in the long term. After Henry's death, the Portuguese caravels would round the southern tip of Africa at the Cape of Good Hope and go on to establish lucrative trading relationships with Africa, India, and China. Later caravels were larger and evolved to include an additional mast with square rigging to improve windward performance. Such ships explored farther expanses of the globe. Most famous among these were the ships with which Christopher Columbus discovered America. The *Niña* and the *Pinta* were lateen-rigged caravels. The *Santa Maria* carried a combination of lateen and square-rigged sails and benefited from several features introduced by the caravel. The caravel's marine engineering coupled with advances in navigation and cartography financed by the royal resources of Henry the Navigator were seminal in bringing about what history books invariably refer to as the West's Age of Discovery.

In contrast to the late medieval setting in which Henry the Navigator worked, James Watt, born in 1736, lived in the increasingly capitalist society of eighteenth-century Great Britain. Not surprisingly, the material resources—the risk capital—that allowed him to engineer the whole series of inventions related to the steam engine came from quite different sources. The steam engine, for which Watt became famous, like the caravel and most other seminal inventions, had strong roots in prior technologies. Its immediate predecessor, the Newcomen engine, had been pumping water from the mines of Cornwall since its first appearance in 1712.

Watt began his professional career close to the University of Glasgow as the proprietor of a shop for making and repairing instruments. There, in his formative years, he benefited greatly from his interactions with the academic community, most particularly from research related to the properties of heat and temperature. One of his tasks entailed attempting to fix a model of a Newcomen engine, intended for use in teaching, so that it would function properly. With his increasing understanding of the nature of heat, it became clear to him not only why the model engine would not mimic the performance of the full-sized machine but also, more importantly, why even the largest Newcomen engines were so inefficient.

The Newcomen engine produced steam in a boiler at slightly above atmospheric pressure. As indicated in figure 17 and detailed in figure 18, the steam filled a vertical cylinder fitted on its top with a piston. After the steam valve from the boiler closed, a jet of cold water condensed the steam and created a partial vacuum. Atmospheric pressure then pushed the piston downward into the evacuated cylinder, creating a power stroke. The downward stroke transmitted power through a chain to a pivoted working beam; the other end of the beam then lifted the pump rod. Following the power stroke, the weight of the pump pulled the piston upward as steam again entered the cylinder, commencing the next cycle. A clever system of water pumps and nonreturn valves sequenced the alternate injection and extraction of steam and water from the cylinder.

The heat losses that resulted in the engine's inefficiency were not understood before Watt. But it was an important problem, for the poor efficiency resulted in vast amounts of coal being transported from northern England's mines to feed the pump boilers of Cornwall, where the engines found their primary use. The loss of efficiency arose from the cycling of Newcomen's steam cylinder between hot and cold, as steam and cold water were alternately injected. Watt's profound realization was that with each quenching to condense the steam, all of the heat was lost from the cylinder walls. Thus with each cycle, the steam entering the chamber had to furnish enough heat to raise once again the temperature of the cylinder walls. This heat was wasted since it produced no useful work.

In 1765, Watt hypothesized a way to circumvent the engine's inefficiency. His concept is compared to Newcomen's in figure 18. Both engines utilized the working beam-and-chain arrangement shown in figure 17 to transmit power from the engines to the pumps they operated. Watt separated the condensation function into a second chamber that was connected to the primary piston cylinder through a pipe. He encased the cylinder in an outer vessel that received steam from the boiler and was insulated from the exterior

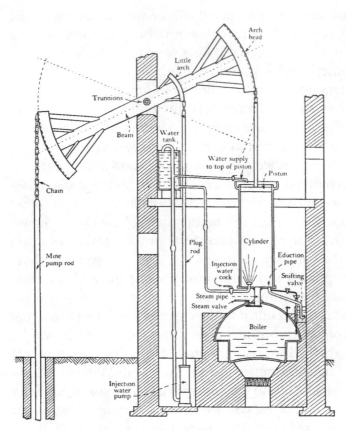

Figure 17. A Newcomen engine. (From H. W. Dickinson, *A Short History of the Steam Engine* [London: Cambridge University Press, 1939], p. 50. Reprinted with the permission of Cambridge University Press.)

environment, allowing the piston and cylinder walls to remain hot at all times. Likewise, the condenser chamber was kept cool by placing it in a container of water, ironically called the "hot well." An air pump removed water and trapped air from the condenser.

Watt's system operated as follows. Steam first filled the cylinder through a valve in the piston head as the piston moved upward. As the piston reached the cylinder's top, the piston head valve closed, and the valve to the condenser opened. This caused steam from the hot cylinder to fill the condensing chamber, where a cold-water jet and the cool walls quickly caused it to condense. The partial vacuum thus created in the chamber sucked the piston downward, providing the power stroke. With the piston head valve then opened, and the

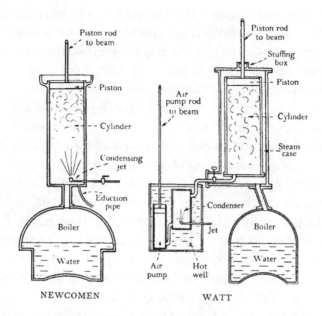

Figure 18. Comparison of Newcomen and Watt steam engines.
(From Dickinson, *A Short History of the Steam Engine*, p. 67.
Reprinted with the permission of Cambridge University Press.)

condenser valve closed, the upward force transmitted through the pivot beam caused the piston again to rise and caused the cylinder once again to fill with steam.

Watt proved with small model prototypes that his separate condenser configurations resulted in much greater efficiencies, and, in 1769, he obtained a patent for the device. The small prototype that embodied his new conceptual design, however, was a long way from a workable industrial engine.

Watt's need to provide for his family did not allow him adequate time to pursue the detailed engineering design and experimentation necessary to turn his concept into reality. Watt's partnership in an instrument-making business had provided him with his livelihood, with modest facilities and at least some amount of time for engine experimentation. Following the death of his partner, however, the business was dissolved. Forced to earn a living by other means, Watt

had little time for engine experimentation. He took up civil engineering and became involved in making surveys for several canals. To
some extent the work was seasonal, so intermittently he continued
his engine development work with borrowed money. But his
resources were insufficient to allow the successful pursuit of his goals.

Prospects for the Watt engine improved dramatically when he
became affiliated with Matthew Boulton. Boulton was a farsighted
industrialist who could envision the potential for Watt's engine. Moreover, Boulton's already successful manufacturing pursuits allowed him
to provide the resources that Watt needed to engineer a steam engine
with a separate condenser to an industrially viable state. In 1775, they
formed the partnership of Boulton and Watt. The enterprise flourished,
lasting until their retirement, when they passed it to their sons.

Together they secured an extension of Watt's patent rights, producing a virtual monopoly on condensing engines for twenty-five
years. Boulton ably managed financial and commercial matters and
provided Watt with laboratory facilities and skilled assistants. Watt
devoted his energies to overcoming design challenges and manufacturing difficulties, particularly in fabricating sufficiently leak-tight
vessels. The partnership met with success in 1777 when they placed
their first commercial pumping engine in service, and it proved to
require only about one-fourth of the coal needed for a Newcomen
engine to pump the same amount of water. Over the succeeding
years, Boulton and Watt engines largely supplanted the less efficient
Newcomen engines, not only for pumping water from mines but
also for city water works and other applications.

Boulton foresaw the power shortage emerging in England's
growing industries as the available sites for waterwheels were being
exhausted. The problem was that the Watt engine, like the Newcomen engine before it, produced a reciprocating motion. The
engines were suitable for pumping, but the rotary motion supplied
by waterwheels was needed for many of the processes in the textile
and other burgeoning industries. So great did the power shortage
become that reciprocating steam engines were sometimes used for
pumping water from a waterwheel's exhaust pond back into its
upstream reservoir for reuse.

Boulton encouraged and cajoled Watt to address this problem. Watt responded by inventing several devices that he combined to achieve an effective rotary steam engine. Indicative of capitalist society, he did not utilize a system based on the most obvious mechanism, the crank, because of potential patent infringement problems. Instead, he developed an elaborate gear mechanism, called a planetary gear, which achieved rotary motion while circumventing the patents held by others. This was but one of a stream of significant innovations. He invented the double-acting engine, in which steam was applied to both sides of the cylinder to double the power per cycle, a paralleled motion linkage to keep the piston rod vertical while driving the rocking beam, and an automatic governor that used feedback to control engine speed.

With Boulton's financial patronage, Watt's technical innovations, and the protection of their patents, their firm became the worldwide masters in the production of steam engines. The Watt engine became the source of power that drove the industrial revolution. For the burning of coal to power steam engines gave humankind a source of energy that far exceeded what had been available from flowing water, from blowing wind, or from the effort of animal muscle.

Little more that a century after Watt's invention came another technological breakthrough, noteworthy not only for its momentous legacy but also for the novel means by which it was financed. In 1896 Orville and Wilbur Wright became seriously interested in aviation and in learning how to fly. To proceed, they had to secure a means to support their program of engineering development. Unlike Watt and many other inventors, they did not seek support from financial backers; instead, they were determined to be self-supporting. Indeed, as their efforts met with success, though they could easily obtain outside support, they refused it in order to maintain their independence. There was no family wealth to contribute to their independence. They were proprietors of a small shop that repaired, sold, and manufactured bicycles in Dayton, Ohio. The Wright bicycle shop was a time-consuming business, with the heaviest work beginning in January and lasting well into the summer, followed by a slacker period from September through December. The Wright brothers carefully husbanded their time and resources to mesh four years of engineering research within the seasonal and financial confines of their small business.

The conceptual design of the fixed-wing airplane had been established nearly a century earlier. Sir George Cayley, an English baronet from Yorkshire, made the intellectual leap from the fanciful flapping-wing school of aviation that had existed since the Greek myth of Daedalus and Icarus to conceive the airplane as we know it. His 1810 publication "On Aerial Navigation" laid out the basic foundations of aerodynamics and flight control: "The whole problem is confined within these limits—to make a surface support a given weight by the application of power to the resistance of air."[2] On a small silver disc commemorating his breakthrough, shown in figure 19, he engraved a fixed-wing aircraft, with a fuselage and a tail structure consisting of vertical and horizontal control surfaces. The force diagram on the opposite side indicated his understanding of lift and drag. He understood that air flowing past an appropriately shaped wing creates an upward pressure to "lift" the aircraft from the ground, while air resistance creates a "drag" that retards the craft's forward motion.

Cayley performed aeronautical experiments and built unmanned gliders early in the nineteenth century. By the end of the nineteenth century, enthusiasts were able to produce gliders with sufficient lift to keep them in the air for more than a few seconds. Moreover, gasoline engines, being developed for automobiles, were growing more powerful while decreasing in weight; soon they would be capable of providing the power that, when added to a glider, would create an airplane. Glider accidents, however, were all too frequent, and the death of Otto Lilienthal, a German famous for his aeronautical experiments, was on the brothers' minds as they studied the aviation literature. They correctly identified the major remaining obstacle to flight: maintaining aircraft stability and being

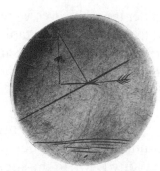

Figure 19. Sir George Cayley's 1799 silver disk. *Left,* design showing wings and tail of an airplane; *right,* force diagram showing lift and drag on a wing surface. (From Peter L. Jakab, *Visions of a Flying Machine* (Washington, DC: Smithsonian Institution Press, 1990], pp. 22, 23.)

able to make turns in flight without losing control. The problem of stability and control was what they set out to solve.

They theorized that they could stabilize an aircraft against wind buffeting by using a wing-warping mechanism that mimicked the behavior that they had observed in birds, particularly in soaring buzzards. Their engineering challenge was first to recreate this mechanism in a glider controllable by the pilot and then to build a powered, controllable aircraft. The brothers' efforts over the next four years synthesized elements of analysis, experiment, design, and testing that were advanced for their day and pointed to modern engineering methodologies.

Since the Wright brothers planned to be their own test pilots, they emphasized safety in their work. They started in 1899 by building a kite, illustrated in figure 20, with the wing characteristics of a glider. They experimented with the control mechanisms by controlling the kite though four lines to the ground. During the busy season at the bicycle shop they found time to design and build a glider. They took their first "vacation" trip in the summer of 1900 to Kitty Hawk, North Carolina, where favorable winds facilitated glider flights. Flying their glider primarily as a kite but then as a manned glider, they gathered enough data and experience to design a larger glider for their return to Kitty Hawk in the summer of 1901.

Their 1901 glider tests, however, were disappointing. They reluctantly concluded that the published tables on lift and drag that they had used to determine the wing shape and area were woefully inaccurate. To remedy this, they built a wind tunnel in their shop that winter and made careful measurements of lift and drag on a variety of different wing shapes and angles. The airfoil data they generated was to serve them well for the next ten years. They combined their new data with a number of other design innovations in the glider that they tested at Kitty Hawk in the summer of 1902. This time the tests were successful. The warped-wing mechanism, when coordinated with a vertical rudder, resulted not only in stable flight but a glider that they could pilot through controlled turns.

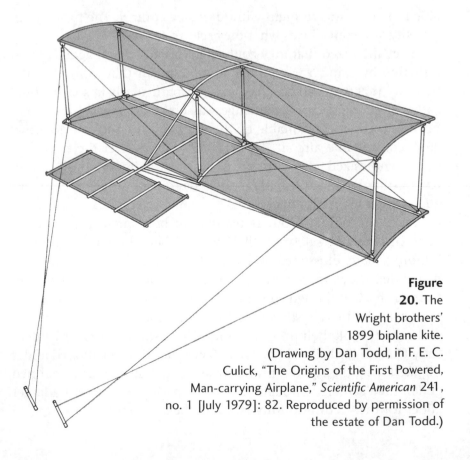

Figure 20. The Wright brothers' 1899 biplane kite. (Drawing by Dan Todd, in F. E. C. Culick, "The Origins of the First Powered, Man-carrying Airplane," *Scientific American* 241, no. 1 [July 1979]: 82. Reproduced by permission of the estate of Dan Todd.)

The final step was to add power. Unable to purchase a suitable engine, they built their own, tailoring it to the needs of flight. In attempting to design a propeller, however, they found that any relevant theory was virtually nonexistent. In what was probably their greatest analytical accomplishment, they developed what would later be known as blade element theory and used it to design and build propellers that were superior to any then in existence. They added the engine and twin propellers to the "Flyer" that they engineered from wood, wire, and fabric. They returned to Kitty Hawk late in the summer and undertook months of effort. Finally, on December 17, 1903, the Wright brothers piloted the first controlled flights of a powered aircraft. The airplane was born.

In financing their engineering activities through their bicycle shop's operation, tremendous demands were placed on their time. Nevertheless, the shop also provided them with several advantages. Their skill and their facilities for making and repairing bicycles were invaluable in constructing the glider components, wind tunnel, engine, and other apparatus with a minimum of cost and delay. The autumn lull in their business enabled the annual trips to Kitty Hawk that allowed them to conduct their own tests. Thus, in addition to being design engineers, they served as test pilots and manufacturing engineers. The intimate feedback they received from all these activities provided a holistic approach compared to those of Octave Chanute, Samuel Pierpont Langley, and other aviation pioneers who designed gliders or aircraft but who had the financial resources to contract others to build them and to hire pilots to fly them. Finally, the brothers were keenly aware of the subtle analogies between cycling, where the rider provides stability and control, as compared to the challenges of stable, controlled flight.

The Wright brothers' experience brings into focus another dimension of the financing of engineering development. By the turn of the twentieth century, science and engineering were becoming increasingly intertwined. The brothers were certainly doing science or at least engineering science when they performed their wind tunnel experiments to improve lift data, when they formulated propeller theory, and in numerous other facets of their quest. But they did not publish their results or make them available to the scientific community; they kept them secret, thereby forgoing the acclaim of the scientific community for years, even after they obtained broad patents to secure ownership of their method for aircraft control.

Comparing the Wright brothers to Chanute and Langley, the two most prominent American aviation pioneers of the time, is instructive. Chanute and Langley both freely published results of their tests and experiments in order to advance the science of aeronautics. But their situations were entirely different. Chanute had already become financially independent as a result of his civil engineering achievements in the railroad industry, whereas Langley was an academic whose research was funded by the Smithsonian Institution and other agencies of government. Their prestige and satisfaction derived not from financial reward but from the widespread recognition of their accomplishments. In contrast, Wilbur and Orville Wright's goal was to leave the bicycle business and secure their financial future in aviation. Thus, even after securing patents for the embodiment of their research, they were reluctant to publicize their scientific data or their engineering methodologies for fear that they would lose their competitive advantage. Indeed, after they founded the Wright Aviation Company, their efforts increasingly moved toward the legal activity surrounding patent infringement suits and away from further engineering innovation. They, like Watt and others both before and after them, were torn by the tensions often found in engineering between the culture of science and that of business.

"This is really high-tech stuff," I thought, as I traveled from my university office to the suburban Chicago site where a meeting was scheduled with the project engineer to discuss our continued collaboration. The project had at its root recent work in materials science that had produced so-called high-temperature superconductors. Superconductors are solids that display near total absence of electrical resistance and the energy dissipation that goes with it. Until recently, however, superconductivity could be achieved only within a few degrees of absolute zero. Because very costly liquid helium cooling systems are required to maintain such low temperatures, superconductors have remained impractical for most applications. The new high-temperature superconducting materials, however, show no electrical resistance at temperatures as high as one hundred degrees Kelvin (i.e., above absolute zero), allowing the use of much cheaper liquid nitrogen cooling systems. The economy of liquid nitrogen cooling and the continually improving properties of super-conductors have made a new range of applications conceivable.

The project focused on a fascinating property of superconduc-tors: the ability to stably levitate magnets so that they make no contact with any material body. Such levitation using small mag-nets had been under increased study in research laboratories. The goal of this project, however, went far beyond the research labora-tory. It was to create an economical energy storage flywheel. Placed in a vacuum chamber to eliminate air resistance, a flywheel rotor equipped with superconducting magnets can spin nearly indefi-nitely, since the absence of physical contact between the wheel and its surroundings eliminates virtually all friction. Fit with motor-generators to convert electrical energy to mechanical energy and back again, a successful flywheel would in fact constitute a mechanical battery, with the energy stored in the spinning rotor. If the flywheel can reach high enough rotational speeds to store suf-ficient energy, and if it can be mass-produced at a reasonable price, then there would be exciting commercial prospects. Electric utili-ties, for example, might use flywheels to store energy produced at night when demand is low by using the electrical motor to accel-erate the wheel to high spin rates. Then during the day, when demand is high, they could extract mechanical energy from the wheel to drive an electric generator.

The work was at an exciting if tension-filled stage. The project team had performed many tabletop experiments to understand the unique features of superconducting levitation. Now spin tests were in progress in larger vacuum chambers to examine the behavior of magnets, fiberglass composite rotors, and other constituents of the proposed flywheel design. Each rotating component needed to withstand the large centrifugal forces generated in the flywheel spinning at very high rates. A great deal of analysis and many computer simulations were necessary to predict stress levels and understand the various vibrations and wobbles that the wheel might develop with ever-higher rates of spin. The problems encountered were interdisciplinary tangles of solid and fracture mechanics, electromagnetic theory, rotor dynamics, heat transfer, and more. The limits of knowledge available to the staff were humbling when compared to what was needed to predict with certainty the operational behavior of an energy storage flywheel.

The project's primary thrust was now in designing and building a full-sized prototype, upon which a commercial wheel would be based. The staff was under time pressure, making design modifications to include new knowledge from experiments and calculations, ordering components for the vacuum and cooling systems, and assembling the motor-generator and its associated power electronics. As with all prototypes, uncertainty still lurked wherever the engineering had outrun the scientific knowledge base and where the data available on the new superconducting materials was not yet sufficient to perform more sophisticated computer simulations. We were all hoping for the best, but precautions had to be taken against the possibility of a catastrophic failure as we proceeded to higher spin rates and stress levels. With so much untested technology, we could not preclude the possibility of the wheel spinning unstably off its axis or disintegrating into high-velocity fragments. As a precaution, the prototype tests would take place in a large cavity below ground level, covered with thick slabs of reinforced concrete.

Arriving at the site, I took a quick look down into the concrete vault to see how preparation of the prototype was coming along, and stopped briefly to chat with a young engineer who was readying some of the components. I walked past the hut where the banks of instruments for recording test data were located and by the video monitors from where visual observations would take place. Entering the project engineer's office, I glanced at the newest schematic drawings, containing the latest flywheel design modifications, and briefly looked at project schedules pasted on the walls, above the laptop computers where CAD drawings filled the screens. In addition to the responsibilities for managing budget and schedule, the project engineer was immersed in details of the array of technical issues and engineering challenges that he faced. His keen attention to technical detail would serve him well in the discussion that I thought we were about to have concerning recent analyses of the rotor dynamics of superconducting bearings.

As our conversation began, however, it became apparent that the project manager was preoccupied with larger issues than rotor dynamics. There were concerns about patent applications and the ability to protect the flywheel design knowledge from competing groups in the United States and abroad. Even more important were budget and schedule concerns. The growth of salary and equipment expenditures with the need for more precise experiments and additional spin tests had become a nagging concern. Delays caused by late arrival of components, complications in perfecting magnet-rotor bonding techniques, and a myriad of other details were causing the schedule to slip. The number and size of subcontracts to produce the superconductors, the composite rotor, and other components were stretching the budget, as were the ongoing rental charges for space and facilities.

The corporate sponsors seemed less excited than the project's engineers and physicists with the breakthrough technology, the new physical phenomena uncovered, or the engineering methodologies developed. Understandably, they focused on the size and certainty of the return on their investment. The project engineer was faced with the task of convincing the corporate sponsors not to lose patience—even if the prototype tests went less than perfectly, the development costs exceeded earlier estimates, or the schedule for beginning manufacture of a commercial wheel had to be extended. He would have to negotiate once again the size of the budget and the length of time that the sponsors would allot to achieving creation of a commercially viable flywheel. Likely, he would need to expend much of his energy in ongoing efforts to secure additional venture capital needed to bring the superconducting flywheel project to fruition.

Driving back to the university, I thought of the relationships between scientific discovery, engineering innovation, and venture capital in the context of the superconducting flywheel. As I began to think of other engineers of other eras, of the Wright brothers, of James Watt, and of Henry the Navigator, I imagined the setting of a medieval monastery, the one with which I was familiar at Hirsau. The project engineer was dressed in the garb of a Benedictine monk as he approached the abbot for an audience. He was about to discuss the efforts that he and his monastic brothers had undertaken to build a waterwheel on the stream that ran past the monastery. His hope was to secure the abbot's patience and convince him to commit more monks' time to the waterpower project, time that could otherwise be spent in agricultural or other pursuits to sustain the monastery. He would have to convince the abbot that the small material surpluses created through the brotherhood's hard labor would best be expended in further pursuit of waterwheel development. He would have to persuade the abbot that in the foreseeable future waterpower would successfully grind grain, full cloth, or perform other tasks that would lessen the burden of physical labor on the monastery's brothers and allow them more time for meditation and scholarship. Returning to the present, I realized that my meeting had not been unique to this high-tech age. We were addressing issues related to technological patronage that are as old as engineering itself.

Five

The Mind's Eye

Visiting any art museum, or for that matter thumbing through any art history book, I never cease to be struck by the momentous transformation in painting style that occurred between the late Middle Ages and the Renaissance. The change from gold-leaf backgrounds to blue sky with white fluffy clouds, from the flat—in some cases almost childlike—figures to the near-photographic realism of human and animal figures, and the change from depthless renderings to three-dimensional perspectives augmented by light and shadowing is truly striking.

Medieval and Renaissance paintings, however, came from entirely different motivations. In the Middle Ages, the stories and deeds portrayed by anonymous painters often were of a theological nature. Their primary focus was on human—frequently biblical—characters, with little attention given to the animals, buildings, or landscapes. Artists and artisans expressed importance as size; God was largest, followed by kings and nobles, with the smallest figure being common people. They employed color not for realism but as a means to express theological symbolism. There was no attempt at making the paintings appear three-dimensional, let alone to include the subtle colorations of natural light.

In contrast, Renaissance painters learned to reflect their subjects as they actually appeared—to create in their paintings windows on the world. Leonardo da Vinci's *Last Supper* offers a good example. Certainly, he imbued the painting with religious symbolism, but it is subtly reflected within the picture's near-photographic accuracy. The figures are lifelike, their humanity revealed through the subtlety of their expressions. The drawing appears three-dimensional, displaying a strong perception of depth. The room is in perfect perspective, and the view through the window reveals the beauty of the Italian landscape. The artist's perception is very different from the medieval frescos of just a century before. Leonardo, as well as other artists, put aside his philosophical and theological preferences and started measuring distances as they actually appeared to the eye, drawing them in proper proportion to one another. Color became a medium for them to portray more accurately the world as it really appeared.

Far more than artistic taste, however, was involved in the transformation of the visual arts. Painters of the Renaissance, Leonardo foremost among them, were precursors of the scientific revolution that came into full bloom a century or more later. It's no surprise that these Renaissance painters have often been considered scientists as well as artists. For their visualization of the world constituted the beginning of observing nature objectively. Keen observation, unobstructed by theological or philosophical preconceptions, was a necessary prerequisite to experimentation and the application of mathematics to the study of nature.

The profusion of carefully drawn illustrations that survive in the many volumes of Leonardo da Vinci's notebooks portray convincingly the role played by objective drawings as a basis for the more quantitative methods of science that were to come much later. His drawings of anatomy show accurate proportions of the human body, often revealing the position and size of internal organs. Likewise, the animals, trees, and flowers in his notebooks appear to be almost alive. And his observations of water—of its flow patterns, tumbling over rocks and swirling in vortices—resemble what might be found in notebooks documenting observations in a present-day hydrodynamics laboratory. The acute observations of Renaissance artists were truly first steps toward more mathematical descriptions of nature by Galileo and Isaac Newton, as well as toward the focused observations that led to Darwin's theory of evolution.

As the term *Renaissance man* has come to mean someone accomplished in both art and science, less recognized is that these painters and acute observers of nature were among the outstanding engineers of their time. Leonardo da Vinci is an excellent case in point. His engineering acumen is reflected both in his career and in the work contained in his proliferation of notebooks. In fact, he was only an occasional painter. The paintings that can be unambiguously attributed to Leonardo are outstanding but quite limited in number. In contrast, he was essentially a practicing engineer and architect for much of his adult life. By following his career in detail we find that he was employed as an engineer in Florence, Milan, and Venice, and later in France, to carry out a variety of civil and military projects. He improved city planning for drainage and flood control, designed bridges, and engineered city and harbor fortifications during a time when the Italian city-states were embroiled in repeated conflicts.

What is more, Leonardo aimed much of his scientific inquiry at providing a rational basis for understanding the engineering challenges of his day. His work was motivated by a desire to replace the trial-and-error approaches that constituted much of engineering practice with more systematic procedures. This search for rational engineering methodologies motivated many of his contributions to the nascent scientific revolution and accounted for many of the illustrations preserved in the thousands of pages of his notebooks. These notebooks reflect his innovative engineering mind, with many pages illustrating devices that he designed: cranes for lifting, gearing arrangements, metalworking machinery, and even visionary ideas for airplanes and helicopters. These are intermingled with sketches of people, plants, buildings, flowing water, and whatever else caught his attention.

The pairing of artist and engineer apparent in Leonardo's notebooks is not one most prevalent in the public's mind. For in this modern age it is science and engineering rather than art and engineering that are the inevitable twins. In the public's mind, the prominent view of the engineer is one who is heavily involved in doing computations and examining the results of experiments and tests. Leonardo's notebooks, however, graphically demonstrate the extent to which drawing predates computation and experimentation in occupying the heart of engineering creativity.

Leonardo's works are only the best-known examples of the many notebooks that survive from prominent engineers of his time. They illustrate both the revolutionary advances in Renaissance drawing and the prominent role of visualization in engineers' thinking. But the importance of drawing and visualization in engineering endeavors did not begin with the Renaissance. What historical records do survive point to an intimate connection between drawing and engineering that extends back to far earlier times.

Unfortunately, very few engineering drawings survive from before the Renaissance. Texts were more easily copied than drawings and survived from antiquity, whereas drawings were corrupted. Thus whereas ten books of text by the Roman architect-engineer Vitruvius survive, all of the illustrations have been lost. Likewise, the guild books in which medieval craftsmen illustrated their methods contained closely guarded trade secrets that were rarely revealed to outsiders. The medieval notebook of Villard de Honnecourt and a few other drawings of the cathedral builders constitute most of what survives from the Middle Ages. Only with the fifteenth-century invention of the printing press and the subsequent development of wood- or copper-cut illustrations was the crucial ability to make exactly reproducible pictorial statements established.

Nevertheless, there are other indications of the importance of drawing and visualization among the engineers of earlier times. These stem mainly from the great construction projects of the past, where drawings were crucial in communicating instructions from designer to worker. In those times, fingertip knowledge may have sufficed for individuals or small groups of craft workers. But the engineering of large buildings, bridges, or ships necessitated more formal means of communication between the master builders and the large numbers of carpenters, stonemasons, or other craft workers at the construction sites. Thus the master designer had to communicate what was to be constructed with enough detail that those working under his direction could carry out their assigned tasks. Greek architects incised precise drawings of the columns, lintels, and other structural members directly in the marble of the temple walls. They used precise geometrical constructs in both full-size representations and scale drawings to provide instructions to the stonemasons for each successive stage of the temple's construction. More than a millennium later, medieval master masons instructed journeymen by inscribing full-size plans for stonework on tracing-house floors. They left much larger instructional drawings on the cathedral walls as well.

The tie of engineering to illustration may date from antiquity, but the new lens for viewing the world brought by the Renaissance revolutionized engineers' drawings every bit as much as it did artists' paintings. The extent of the upheaval becomes apparent by comparing the sawmill illustration in figure 11, taken from Villard de Honnecourt's medieval notebook, referred to in chapter 3, to a late Renaissance illustration of a similar sawmill shown in figure 21. Villard's drawing is totally lacking in three-dimensionality and as a result suffers in its ability to convey in a clear manner how the saw system works. Figure 21 is more easily understood, for the machinery and human appear drawn in realistic perspective. The revolution in drawing set into motion far-flung changes in the conception and communication of technical ideas. Taken together with the impact of the printing press, it ranks with the scientific and industrial revolutions in the significance of its effects on the practice of engineering today.

Pictorial perspective is arguably the Renaissance's most significant contribution to modern graphics. Evidence abounds that the Greeks understood perspective, but no drawings survived; thus it is the techniques originated by Renaissance artists that have been passed to the present. Perspective provides a uniform convention for the representation of three-dimensional objects that can be readily interpreted. Perspective drawings first appeared around 1425. About a decade later Leon Battista Alberti set down the necessary rules and became known for his system for constructing perspective drawings. His system also included the apparatus shown in figure 22. Note

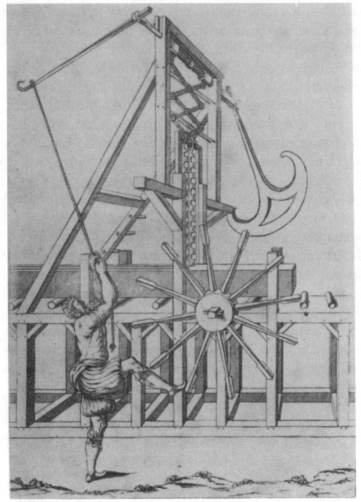

Figure 21. Manually driven saw from Jacques Besson, *Théâtre des instruments mathéma- tiques et méchaniques* (Lyon, 1578).

that the painter views the subject from a fixed eyepiece directly in front of his eye. Between the artist and the subject is a vertical glass window, called Alberti's window, with equally spaced grid lines drawn on it. The horizontal sheet upon which the artist is drawing also contains equally spaced grid lines. Thus, by observing the subject's outline in relationship to the grid lines on the window, he can transfer the image to the horizontal sheet without distortion and therefore obtain a drawing in realistic perspective.

Other forms of drawing that have since proven to be of great

Figure 22. Apparatus for making perspective drawings.
(From Albrecht Dürer, *Hierin sind begriffen vier Bücher*, 1528.)

value to engineers closely followed the use of pictorial perspective. Cutaway drawings, such as the one shown in figure 23 from one of Leonardo's many notebooks, allowed the viewer to look inside and see in a meaningful way the inner working of mechanical apparatus, or for that matter of the human body. They allowed as much of a casing, skin, or other obstruction to be removed as necessary, enabling the observer to see the critical elements of the otherwise hidden mechanism. A second new form was the exploded view drawing. As illustrated in figure 24, again from a Leonardo notebook, these drawings spread out the parts of a machine along a common axis. The result is a more detailed view of the individual parts and a comprehensive picture of how the pieces fit together.

Cutaway and exploded-view drawings allowed Renaissance engineers to portray devices, mechanisms, and structures in rigorous yet imaginative ways to enhance viewers' understanding. Perhaps the new form of drawing that was to have the greatest impact on engineering over the coming centuries, however, was the orthographic projection technique. German Renaissance artist and engineer Albrecht Dürer devised orthographic projections in his efforts to accurately represent the human body. An example of his work from 1528 appears in figure 25. In it he shows three views, at right angles to one another, of a human foot. The inherent ability of such three-view drawings to describe three-dimensional objects unambiguously on flat paper led engineers to adopt the method more than any other for the description of structures and machinery.

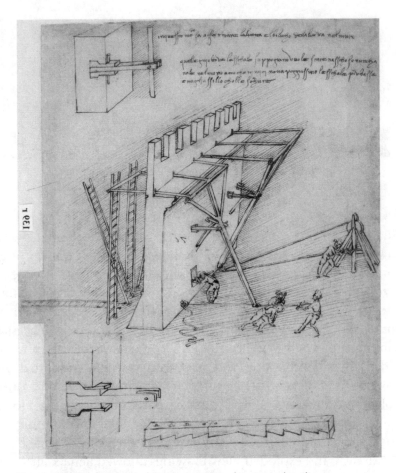

Figure 23. Leonardo da Vinci's cutaway drawings detailing apparatus
for securing horizontal beams to a fortification wall.
(Reproduced by permission of the Ambrosiana Library, Milan.)

But why were new forms of drawing so important to the development of engineering and to the creation of new technologies? In short, they greatly advanced the ability of the engineer to think on paper. In doing so, they allowed the process of conceptual design to be separated from that of actually producing an artifact. This contrasts sharply with the process by which traditional craftworkers produced technological artifacts.

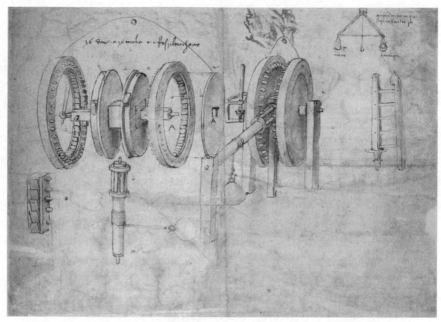

Figure 24. Leonardo da Vinci's drawing of a racket mechanism
in normal (left) and exploded (right) views.
(Reproduced by permission of the Ambrosiana Library, Milan.)

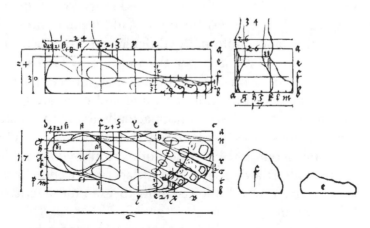

Figure 25. Albrecht Dürer's orthographic projection of a foot.
(From Dürer, *Hierin sind begriffen vier Bücher*, 1528.)

Design and production in the craft process are one and the same. Skilled workers crafted their designs simultaneously as they shaped and assembled an artifact; they didn't first draw and then build them. A self-employed craftworker, or one working in a small shop, had little need for plans or illustrations. A wainwright, for example, built a wagon directly from his own knowledge, gathered during his apprenticeship and through long experience. The wainwright had what is sometimes referred to as fingertip knowledge. This was possible because the wagon differed very little from the many that he had built previously. Likewise, his customer, likely a farmer, already had a good idea of what that wagon should look like, and the farmer could easily point to the small changes that he would want to see made. Thus where design and production process were one and the same—and changed very little from wagon to wagon—the wainwright had no need to set out his plans on paper. At most, he might sketch directly on the wooden planks to show an apprentice the length of a cut or the positioning of a hole.

But the craftworkers' approach was entirely different from what we find in the many volumes of Leonardo's notebooks. Leonardo was conceptualizing new artifacts, mechanisms, and systems. Some of his conceptions were nothing close to anything that had ever been built or even conceived. And there the contrast to the craft worker is striking. For if the craftsman wanted to try something new or different, his only option was to try to build it directly. But this was very limiting and profoundly different in approach from Leonardo, who could think on paper. With the ability to express his three-dimensional ideas on a two-dimensional surface, he could separate design from construction; examine and compare alternatives; correct mistakes and flaws without actually trying to construct the artifact; and adjust sizes, proportions, and shapes as he thought ever more deeply about how his conceptions would work and how they would be built. Drawing separated the act of conceptualization from that of execution and allowed him to examine alternatives without actually building them.

The page upon page of devices sketched in Leonardo's notebooks, along with those found in the notebooks of other Renaissance engineers, expressed a kind of technological exuberance, the mind's creative ability to visualize new and varied technological possibilities. Some of the drawings were visionary, such as Leonardo's portrayals of helicopters and flying machines. Others centered on more mundane challenges such as improving the gearing of machines, reducing the friction in bearings, and improving the method for cutting screws. Of the many devices and systems sketched in Leonardo's notebooks, it is doubtful that many were ever built, and, in some cases, analysis has shown that unresolved conceptual difficulties would have prevented his machines from working. More frequently, however, the nascent state of science and the limitations of Renaissance manufacturing capability presented larger obstacles to their realization.

Leonardo's contemporaries expressed great respect for abstract mathematics and a faith that it would become a valuable tool in engineering. But their notebooks give no indication that they used it much in their professional lives—that possibility was precluded by the primitive state of Renaissance science. Leonardo struggled mightily to transcend these limitations. His inquiring mind led him to make important inroads into the development of the science of engineering. He analyzed force balances, for example, and performed experiments to understand the breaking strength of wires with different diameters and of different lengths. Leonardo began to formulate stress concepts, but he didn't have the mathematical apparatus to come up with a method for predicting, for example, how thick his beams must be not to break or bend unacceptably under loading. It would be a hundred years later before Galileo and his successors would make real progress on that problem. Similarly, the limitations of Renaissance knowledge did not allow him to estimate the degree to which friction would impair the motions of moving parts. His efforts notwithstanding, the state of science was far too rudimentary to be of much value in analyzing the machines that he conceived.

No doubt Leonardo also found frustration in the severe restrictions that the limited materials and manufacturing know-how of the time placed on his ability to turn conceptual designs into reality.

Iron was a crude but valued commodity, too expensive to use for more than small parts of clocks, gunlocks, and other mechanisms, or for the armor and cannon, which the military could afford. For the most part he had only wood in ample supply from which to construct his machines. But even the most skilled craftsmen could not form wood to the precise dimensions and intricate interlocking shapes that his ingenious inventions required, since wood swelled and warped and cracked. His gearing cried out for metal parts whose fabrication was beyond the capability of the blacksmiths of the time. It was difficult enough to shape gear teeth to the precise geometry that he called for, and even if artisans could make such teeth from wood, weakness would cause them to quickly break or wear away. Successful realization of many of Leonardo's conceptions would have to wait for the availability of cheap iron components and the advances of metalworking methods that came three centuries later with the industrial revolution.

The severe confines placed on Renaissance engineers by lack of scientific understanding and manufacturing know-how, however, should not detract from their accomplishments. In synthesizing art and engineering they learned to draw in ways that allowed them to conceive of new technologies by visual brainstorming. Today, as during the Renaissance, engineers are limited by the current state of scientific understanding and manufacturing capability as they deal with increasingly advanced technologies. Still, orthographic projections, realistic perspective, and cutaway and exploded-view drawing have greatly enriched their conceptual capabilities. For thanks to Leonardo and his contemporaries, today's engineers may work out ideas on paper—or by extension, on computer drafting programs—before they attempt to realize them in metals, ceramics, polymers, or stone.

From the late Middle Ages to the present, a trail of sketches in diaries, notebooks, and on the proverbial backs of envelopes documents the imaginations of engineers and architects and their new conceptions. Their minds work in visual as opposed to verbal modes to arrange the three-dimensional logic and workings of new technologies. This occurs not only in structures and mechanical devices where the importance of size, shape, and interlocking arrangements are most obvious but also in more abstract pursuits, where geometry might appear to play less of a role. Thus we may find computer hardware designers sketching alternative configurations of circuitry for future microprocessor chips and software engineers creating diagrams to capture the logical essence of planned computer code. Before mathematical analysis or experimental tests can be performed, before the processes can be worked out for fabrication, assembly, or construction, new technologies must be imagined. And engineers and architects from Leonardo's time to the present invariably first express their imaginations not with words but as sketches, diagrams, and drawings.

In the mid-fifteenth century, the value of the advances in Renaissance drawing was greatly enhanced by the invention of the printing press. The press created an opportunity for engineers to communicate technological ideas to vastly wider audiences. Copper plates and wood carvings allowed the early presses to combine engineering drawing and diagrams with the printed word in volumes that would fill essential functions. Beautiful woodcut illustrations in the Gutenberg Bible and other volumes of the time served to illuminate those texts. But engineering drawings were yet more central to technological writings, and, over time, they would become the lifeblood of engineering communications.

Following the invention of the printing press, drawings of engineers' new and novel ideas for grain mills, cranes, sawmills, and many other elaborate mechanical devices filled heavily illustrated books called "theaters of machines." Many of the machines were visionary, performing tasks and solving mechanical problems that far exceeded the needs or demands of their time. The extravagant machines that filled the pages were, rather, technological possibilities, expressions of the engineers' exuberance and the Renaissance fascination with mechanical contrivances. Some of the machines, of course, did address pressing practical problems. But even most of those would have to wait centuries before manufacturing capabilities and scientific understanding would advance sufficiently to bring their promise to fruition.

The printing press also brought about the spread of a second, quite different class of technical publications: very practical books illustrated the best technological practices of the time. They related the construction of machinery and the know-how of existing industrial processes, allowing readers to replicate technologies as they used the volumes as instruction manuals at distant locations. Figure 26 comes from perhaps the best-known and most beautifully illus-

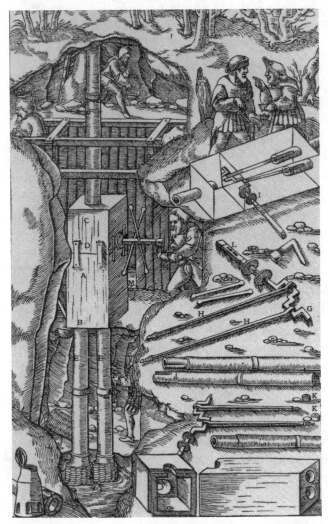

Figure 26. Cutaway view of a manually operated reciprocating water pump. (From Georg Agricola, *De Re Metallica*, 1556. Dover Pictorial Archive Series.)

trated of such books, Georg Agricola's 1556 volume *De Re Metallica*, a treatise on the mining, smelting, and refining of metals and other minerals. Agricola and others filled their books with solid technical information detailing the best practices of the time. They contained numerous illustrations and provided extensive and detailed descriptions of various industrial processes.

From these beginnings, the uses of engineering drawings continued to expand through the ensuing centuries. They presented technological concepts and proposed projects to customers and clients in attractive and understandable formats that served to attract their patronage. They translated design concepts into plans so detailed that workers at remote locations would use them to fabricate pieces and assemble parts into three-dimensional reality. How-to illustrations not only showed manufacturers how they were to build, fabricate, construct, or assemble technological artifacts; they also showed owners and operators how to use, maintain, and repair them. They explained not only how machinery looks but also how it works. Drawing techniques for these purposes have undergone centuries of evolution from before the Renaissance to the present day, an evolution that parallels the industrial development of the West.

A quite different tool for visualization also gained prominence among Renaissance engineers, one that is closer to sculpture than to painting. Three-dimensional scale models have often proven as valuable as drawings in architects' and engineers' work. During the Renaissance, formal competitions often determined the choice of an architect-engineer to carry out monumental building projects, and the winning presentations often rested on the use of a scale model. Brunelleschi's elaborate scale model, for example, won him the commission to build the dome of the cathedral in Florence. Models, not just of buildings but also of fortifications and indeed entire fortified towns, found increasing use as time went on. Later, during the reign of Louis XIV, scale models, often fifteen or twenty feet on a side, provided army officers with a means for studying siege and defense strategies. Perhaps the modeling tradition is strongest, however, in naval architecture, where for centuries both the design and building of great sailing ships relied heavily on the use of scale models.

Shipbuilding offers particular insight into the importance of visualization, for the drawings and modeling techniques of the ship-wrights' tradition extended well into the nineteenth century, changing little until wooden sailing ships gave way to those fabricated from iron and propelled by steam. Although shipwrights used many variations of the process,[1] it invariably began with a scale model of the hull's shape. From laminated layers of wood, a ship's designer would chisel, carve, and sandpaper a three-dimensional likeness of one-half of the vessel's hull, that is, a model of the hull sliced vertically down the centerline from bow to stern. Discussions then followed among the ship's designer and builder; the owner, who was commissioning the construction; and often the captain who would set sail on the ship's maiden voyage. The discussions focused intensely on the model, for its details were all important; the shape of the hull determined both the ship's carrying capacity and its sailing characteristics. Scrutiny went well beyond visual inspection: the participants drew their experienced fingertips over the model's surface to examine contour details that could significantly affect the ship's seaworthiness and economy. As discussions continued, the builder refined the shape of the half-hull with further chiseling, filing, and sanding until the group reached consensus. The finalized model then served as a contractual specification for the vessel's construction.

The shipbuilder used the model hull in a procedure such as that illustrated in figure 27 to guide the ship's construction. With straightedge, calipers, and square he first transferred many exact measurements from the model, often separating the model into its laminated horizontal sections to facilitate the task. From these measurements he laid out on paper a complete geometrical description of the hull's shape, drawn as vertical and horizontal sections of the hull. To begin the vessel's construction, he scaled these sections up to full size by drawing with chalk on the floor of a large shed or other workplace. These full-sized drawings then allowed the ship-

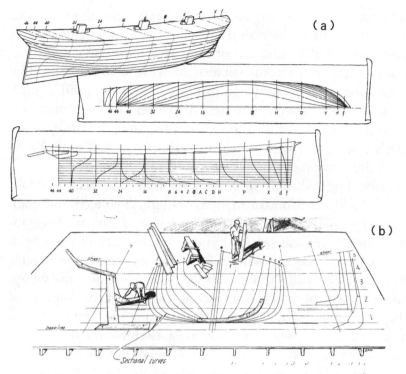

Figure 27. Wooden ship design and construction: *a*, plans made from half-hull scale model; *b*, construction from full-sized loft floor drawings. (Drawings by Sam Manning, in Basil Greenhill, *The Evolution of the Wooden Ship* [New York: Facts on File, 1988], pp. 207, 97. Reproduced by permission.)

wrights to make patterns for the keel, ribs, and other structural members from which they then constructed the vessel's skeleton.

The importance of scale models often extended beyond the process of design and construction. In shipbuilding, as elsewhere, they played a central role in presentations to patrons and clients. In seventeenth-century England, for instance, the Admiralty inspected a detailed scale model not just of the hull but of the entire ship with all of its rigging before giving approval for construction to begin. The large collection of these models preserved in the British Maritime Museum is testament to both the extent of the practice and the exquisite detail that the Admiralty required of them. These models are often cutaways, with only half of the ship remaining in order to reveal details of the construction and the interior arrangements of the lower decks, bulkheads, and ribbing. In a sense these are three-dimensional renditions of the cutaway drawing techniques used by Leonardo. The models allowed admirals to walk around a peak behind the corners. They could view the vessel's likeness from different vantage points to gain a more complete understanding of the designer's intent, and with much less effort than was required to pore over the many pages of drawings and specifications

The modeling tradition continues. Even today, both scale and full-size models and mock-ups of many types serve architects and engineers and their patrons in visualizing buildings, dams, bridges, submarines, aircraft, nuclear reactors, and virtually every other type of technological undertaking. With exactitude, they specify the streamlined shapes of ships' hulls, automobiles' bodies, and aircrafts' wings. They also make explicit the construction details for the framing of buildings, the piping of chemical plants and the interconnections of machine parts. They allow those who must assemble the parts into a unified whole to understand better the sequence in which they must perform their tasks. Present-day designers, engineers, architects, and corporate managers owe much to the heritage of scale models. Far more than two-dimensional drawings, and far more than the written word, three-dimensional models frequently provide engineers with the best means for visual understanding of the complex shapes and geometrical configurations with which they must deal.

During the eighteenth century, the industrial revolution caused rapid acceleration in both the use and the sophistication of engineering drawings. The marked increase in separation of design from manufacture and the replacement of wood and stone by metals as the primary materials for the construction of machinery and structures drove these trends forward. The development of the steam engine, and the central role played by James Watt, discussed earlier, illustrates these changes.

Watt made excellent drawings, evolving the practice rapidly as it became an essential means of communication. Early steam engines were very large, and the detailed layout of the engine house was all-important, because its architectural characteristics had to support and accommodate the engine's heavy iron vessel, its rocker arm, and other large mechanical components. Detailed drawings of each mechanical component were also essential. Watt worked on the conceptual and detailed designs in the city of Birmingham, far removed from the widely scattered sites where the engines found use. Moreover, subcontractors fabricated the various engine components—the vessel, valves, and the mechanical linkages, for example—in yet other cities and towns. These had to be built specifically according to Watt's drawings. Otherwise the construction crews at the engine site would not be able to fit them together, assemble the engine in the newly constructed engine house, and make the whole system function as it should.

During the same time period, drawing methods passed down from masons, carpenters, engravers, and other craftsmen—as well as from the Renaissance engineers—became codified and put on a more mathematical footing. In France, Gaspard Monge published the system of descriptive geometry in 1800 that became a large part of French technical education. It also formed a theoretical foundation for the graphical methods by which present-day engineers give expression to their thoughts. In England, twenty-two years later, William Farish published a treatise on the method of isometric projection. As the comparison in figure 28 indicates, isometric projection, which utilizes parallel lines, is faster and simpler to employ than perspective drawing. Yet, with only slight distortion, it provides comparable ease in visualization. Seeing the object from the orthographic projections, which are most commonly used in engineering drawings, required more effort.

The need for unambiguous specifications—mainly as a profusion of detailed drawings—has become more intense as technological artifacts have become more and more complex. Both the intricacies of individual parts and the number of parts, components, and subsystems in technologies of growing sophistication reflect that

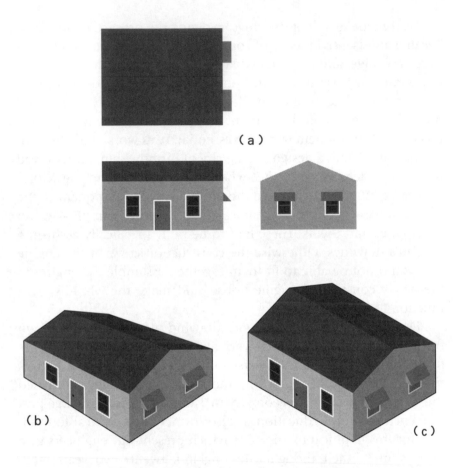

Figure 28. Comparison of three methods of graphic design:
a, orthographic projection; *b*, perspective drawing; *c*, isometric projection.
(Drawings by Michael Epstein.)

complexity. Whether for a single large project, such as a bridge, a
building, or an off-shore oil platform, or for a mass-produced artifact, such as a home appliance or an automobile, the pages of
drawing needed to specify the design and to communicate instructions for parts fabrication and system assembly becomes voluminous. Those for an aircraft, for example, may number in the thousands of pages. The amount of information communicated by these
drawings dwarfs that of Watt's day. Webs of contractors, subcontrac-

tors, and suppliers scattered in cities across the country—and increasingly across the world—fabricate and assemble increasingly large numbers of components, subcomponents, pieces, and parts that must fit seamlessly together to make up the many technologically complex products that occupy our world.

Engineering drawings must specify more than actual size, shape, and material composition. By word, symbol, or number, each dimension or material specification on a drawing must be accompanied by a tolerance, an indication of how close to the ideal value each characteristic must be. For no object's dimensions can be replicated exactly, down to the nearest atom. Likewise, the fractions of the varied constituents that make up an alloyed metal, a ceramic, or polymer cannot be controlled exactly to zero deviation. An automobile axle cannot be economically manufactured to have a diameter of $5.000000000 \cdots$ inches, nor could the steel from which it is made be guaranteed to have a carbon content of exactly $2.000000000 \cdots$ percent. Each of the many lengths, radii, compositions, or other specifications appearing on an engineering drawing must have a tolerance—the so-called plus or minus values—associated with it.

Tolerances are as much a part of engineers' designs as the quantities themselves. Tolerances that are too loose lead to parts that fit together poorly and to the malfunctions and failures that characterize poor-quality products. Tolerances that are too tight lead to skyrocketing manufacturing costs and exorbitant product prices. The quest for tighter tolerances without intolerable increases in production costs is as much with us today as it was in the eighteenth-century shop of James Watt. For just as Watt struggled to obtain iron vessels with clearances within tenths of an inch to reduce steam leakage, today's computer chip designers work to inscribe silicon with features whose dimensions must be accurate to within millionths of an inch. The necessity of including tolerances adds greatly to the information that engineering drawings must carry, drawings that already are likely to be exceedingly complex.

The number of characteristics and their associated tolerances that engineering drawings must detail grows with the complexity of engineered artifacts. The proliferation of drawings required to detail such complexity has driven engineers to use more abstract symbolism to compact their visual representations. Consider drawing at the most detailed level, for example, the rendition of individual mechanical parts. Such parts often contain holes, threading, or other features that are repeated many times. Moreover, designers invariably specify repeated features in standard, identical sizes to reduce fabrication costs and facilitate the use of mass-produced bolts, bearings, and other fittings. Economy in drawing is then achieved by representing these features with simplified symbolic forms. For example, figure 29 shows six bolt thread representations: in the first pair the threading appears in detail, whereas the second and third are successive abstractions in which symbolic notation results in more efficient representations.

As the numbers as well as the complexity of parts and components multiply, the need grows to represent each of them more compactly—and thus abstractly—so that systems with many components can be described on a single page. The parts and components must be represented such that those examining the drawing can understand the logic of their arrangement—how they connect and interact. The viewer must be able to see the forest of the system through the trees of the parts. Thus it is the case, for example, that circuit designers do not draw detailed pictures of resistors, transistors, capacitors, amplifiers, rectifiers, or cables; rather, they represent them symbolically as in figure 30 in order to concentrate on the architecture of the circuitry. Likewise, fluids systems designers do not

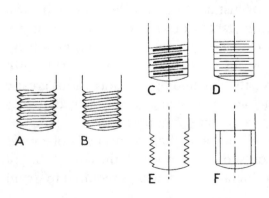

Figure 29. Progressive simplifications in representing a screw thread. (From Peter Jeffrey Booker, *A History of Engineering Drawing* [London: Northgate, 1979], p. 182. Reproduced with the permission of the British Standards Institution under license number 2003SK/181.)

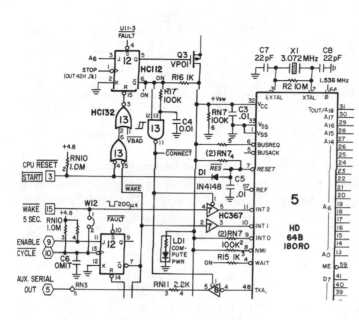

Figure 30. Electronic circuitry diagram. (From Paul Horowitz and Winfield Hill, *The Art of Electronics*, 2nd ed. [Cambridge: Cambridge University Press, 1989], p. 843. Reproduced with the permission of Cambridge University Press.)

represent each pump, motor, valve, and flow regulator in detail. They represent them symbolically, as in figure 31, to compact the drawing and better represent the system's logic.

Whatever their engineering disciplines, practitioners employ highly abstract notations specific to their field. They combine symbols and numbers to represent unambiguously the characteristics of components, dimensions, tolerances, and materials essential to complex designs. Even then, they must accompany their drawings with lists of standardized specifications to keep the drawing from becoming unnecessarily cluttered. As technology advances, layer upon layer of such drawings must be combined in cross-disciplinary hierarchies to provide the comprehensive descriptions necessary to create the technological products and systems to which we are accustomed.

Engineers' drawings and specifications form a pervasive means of communication. For at each level, the pieces of the system, subsystem, or component may not be built locally but purchased from outside suppliers. These suppliers may manufacture the subsystems,

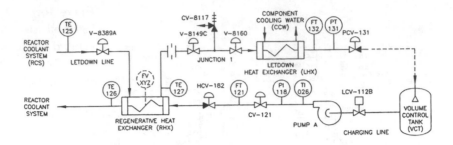

Figure 31. Thermal-fluid system diagram. (From Jacques Reifman and Thomas Y. C. Wei, "PRODIAG, A Process-Independent Transient Diagnostic System—1: Theoretical Concepts," *Nuclear Science and Engineering* 131 [March 1999]: 331. Copyright 1999 by the American Nuclear Society, La Grange Park, Illinois.)

components, or parts in different cities or even different parts of the world, requiring precise communications to make the whole system work. These communications take the form of drawings far more often than written descriptions, and increasingly the symbols on such drawings are standardized throughout the world. Whether their discipline is electrical, mechanical, chemical, or civil, experienced engineers visualize the function of systems in terms of the architecture of components, each displayed in highly abstract form. They think and communicate through the symbolic languages of their technical discipline.

Engineering, of course, is not the sole endeavor that communicates primarily by nonverbal means or that has evolved to employ highly abstract forms of notations to document its creations. Think of the score for a Bach fugue, a Beethoven symphony, or a Verdi opera. Each has a profusion of abstract symbols, of notes and rests, sharps and flats, bass and treble clef signs, accents and other markings streaming across page after page of staff paper. It is arguable that the creative design processes also bear some similarities. Imagine Leonardo, with his sketchbook, drawing a machine or structure, trying alternatives, modifying concepts, and making judgments in his mind's eye as to what will work best without needing to build the artifact. Likewise, imagine Beethoven with his staff paper, sketching a symphonic passage, examining and discarding alternatives, and portraying symbolically the architecture of his musical conception, long before performance. Later in the process, conceptual sketches must be filled out and codified as detailed designs to provide a comprehensive set of instructions for production or performance. Before the symphony could debut, Beethoven had to write a detailed score including each of the instruments in the orchestra. Other musicians, charged with performing the work, must understand that symbolism. They must decipher the themes, rhythms, and sonorities to hear the composer's artistic intent. Likewise, in modern times, technological conceptions must be translated into detailed designs, documented as sets of working drawings. And the engineers, technicians, and others charged with implementing the idea to create an artifact must understand the symbolism pervasive in these drawings.

Today's engineer, as she sits in front of a computer more powerful than she could have dreamed of having even a few years ago, has at her fingertips the accumulation of centuries of engineering knowledge and know-how. She is the beneficiary of the genius of Renaissance engineers' and artists' revolutionary contributions to drawing as a conceptual design. Her software rests on the scientific understanding of engineered artifacts that has taken great strides forward in the centuries since Galileo and today accumulates at an ever-increasing pace. She is also the beneficiary of manufacturing technology that can mass-produce her designs with levels of quality and economy that were inconceivable in earlier times.

Her system embodies the rapid advances in computing and networking achieved over the last decade. They have revolutionized her ability to conceptualize and likewise to communicate and analyze engineering innovations. Large color monitors attached to powerful computing platforms and to networks that stretch throughout corporate offices and increasingly across the world have replaced the drawing boards that were once the center of engineering design activity. Engineers still sketch and think on paper before formalizing their ideas in computerized drawing. But as surely as writers put aside their pens and typewriters to compose with a computer word processor, increasingly, engineers use a mouse or electromagnetic pen to conceptualize their thoughts directly on the computer screen, relying on paper less and less.

Advantages accrue immediately in committing thoughts to the computer instead of paper. Our engineer can use computerized solid-modeling techniques to detail an artifact's shape and size, and she can quickly switch between orthographic projections and perspective drawings to view her work. She can vary the perspective's point of vision with the movement of a mouse, allowing her to walk around and inspect her creation virtually from many vantage points or to home in on a particular part or component for more detailed examination. With a few keystrokes, advanced software provides her with cutaway drawings to reveal internal workings of a mechanism from any desired angle or exploded drawings to make clear how its parts fit together. Automated shading and shadow add further realism, and animation indicates how the parts move relative to one another. This realism is central to the increasing efficiency with which she can conceptualize designs. The computer's power adds to the speed with which she can examine, evaluate, adopt, or discard many alternatives and variations of a design. But equally important are the advances that powerful computing capability brings to communication and analysis.

Once she has created a virtual design of a an artifact or system on her computer, she can communicate it through the Internet to others who need it: to the engineers who must design mating parts, to the suppliers who will provide standardized components, and to the manufacturing process engineers who must design and build the tools and lay out the industrial plant to mass-produce it. Computers facilitate rapid communication of inevitable design changes and modifications between these groups as they collaborate to optimize performance, improve manufacturability, and reduce costs. Vanished is much of the delay and confusion that frequently occurred when many pages of engineering drawings had to be redone, distributed, and coordinated as the result of seemingly minor changes in dimensions, finishes, or other details.

The computer also brings to bear the accumulated understanding of the scientific laws underlying the behavior of engineered objects—the reaction chemistry, solid and fluid mechanics, control and electromagnetic theory, and other engineering science—to which she has access. Design software increasingly is coupled to computer codes utilizing powerful computational techniques to determine the stresses, temperatures, voltages, or reaction rates resulting from the designers' decisions. Three-dimensional displays of the computed results then indicate through color where stress concentrations, hot spots, or electrical overloads occur. Such computer-generated graphics are valuable tools, assisting her in determining where more material, cooling, or electrical insulation is needed or, conversely, where it can be safely eliminated to reduce costs.

Yet, as she utilizes her powerful workstation, our engineer must be keenly aware that it does have limitations. The computer-aided designs that she generates can be no better than her creative imagination and her ability to relate the characteristics of engineered artifacts to human needs and wants. The analyses she performs can be no better than the scientific understanding that went into the creation of the software. Invariably, innovative new technology outruns its scientific base. Unstudied phenomena or those that are not completely understood may underlie her creation's behavior, and her design may be exposed to conditions of use or environment not adequately modeled by the simulation software. Thus her judgment is critical in determining what experiments must augment computer analysis and the extent to which verification testing must confirm the computer's predictions. She must also understand manufacturing capabilities. For drawing something on the computer doesn't necessarily mean that it can be built, and whether in semiconductors or superconductors, advancing designs frequently probe the fringes of what can be mass-produced.

These caveats notwithstanding, our engineer brings to her work the proud heritage of Leonardo and his Renaissance colleagues as she creates visual images on her workstation screen. Leonardo also must have been cognizant that sketching ingenious devices did not provide him with the scientific knowledge to ensure that they would work. He must have been all too aware that the carpenters and blacksmiths of his time were incapable of building many of his creations. But our engineer, in committing her visual conceptions of new technological artifacts to the computer, follows in his footsteps and participates in a venerable tradition of her profession. For as argued eloquently by the historian Eugene Ferguson, engineering concepts originate in the mind's eye.[2] More than computed numbers, even more than the written word, it is visualization that is fundamental to creating new technology and to communicating these creations to other engineers, as well as to clients, patrons, and the general public.

Six

All Things in Measure, Number, and Weight

Visiting college campuses has long been among the most
enjoyable aspects of my travels. The University of Cambridge
is a favorite, for it is the quintessential college town, with its cobblestone streets, Gothic architecture, and renowned college choirs, and
where even the street musicians sometimes perform as string quar-
tets. During a recent visit, my wife and I lingered over an early lunch
at Eagle Pub, reflecting on the previous afternoon's experience. Only
the sun's fading rays, hued as they entered through vast expanses of
stained glass, and the points of candlelight emanating from the cho-
risters' lecterns had illuminated the intricate fan vaulting high above
us in the famed chapel of King's College. The visual artistry of the
setting had been in unforgettable harmony with the choir's sung
service of ancient psalms and hymns, and with the Bach fugue with
which the chapel organ brought the Evensong to a close.

But our activities that day were to be focused on science, and
where better to begin than at the pub where Francis Crick and James
Watson conferred frequently during their quest for the structure of
DNA? We left the pub and walked across Cambridge, past the University Senate House. There C. P. Snow had delivered his historic "Two
Cultures" lecture, which called attention to the widening cultural gap
between the sciences and the humanities. We passed Trinity College,
where Newton once taught, and walked by the house were Darwin

had lived. I felt as if we had been strolling through the history of science itself as we approached the edge of town to tour the modern facilities of the university's Cavendish Laboratory.

Entering, we were greeted by the portraits of famed Cavendish professors: James Clerk Maxwell, Lord Raleigh, J. J. Thomson, Ernest Rutherford, Lawrence Bragg, and others whose names appear so prominently in the physics and engineering textbooks of today. Experimental apparatus used by Cavendish faculty in making their seminal contributions were also preserved in the laboratory's front lobby. The gas discharge tube with which Thomson discovered the electron little more than a century ago seemed particularly apropos to our visit, since we were headed for the Microelectronics Research Center.

Amid electron-beam and photolithography facilities, electron microscopes, and other equipment, students and faculty pursued experiments and theory that were far removed from everyday experience. They were studying nanostructures with dimensions measured in billionths of a meter, so small that they could not be seen with ordinary microscopes, no matter how powerful. Nanowires, quantum dots, nanopillars, and similar structures they were working with had dimensions that sometimes amount to only a dozen or so atoms placed end-to-end, and the investigators could view them only with the most powerful scanning electron microscopes. Fabricating such devices and observing them was difficult, but understanding their behavior is even more of a challenge. At such minuscule scales, the wave-particle duality of quantum mechanics manifests itself in phenomena that have no analogue in the everyday world of our experience. The appearance of quantum tunneling, coulomb blockade, and other phenomena unique to the nanoscale makes explaining experiments on these tiny aggregates of atoms exceedingly difficult but also raises exciting prospects for discovery.

The work was on the frontier of scientific discovery, and yet it was also indicative of the close amalgamation of science and engineering brought together by the twentieth century. As investigators attempted to understand their devices' behavior, they also pondered how they might eventually be put to practical use. As they used powerful chemicals and advanced lithographic techniques to control the heights of the pillars and the diameters of the dots that they fabricated, thoughts crossed their minds concerning how such devices, once a use was identified, might be fabricated cheaply in large numbers. Only then could a potential application pass from laboratory curiosity to affordable product, perhaps a smaller, faster computing device, or a probe or sensor for biological or medical application. The laboratories we visited exemplified the close intertwining of science, engineering design, and manufacturing that characterizes much of modern technology.

The modern synergy of science and engineering may lead one to the temptation to equate the two. But suppose we look back before Newton's laws, the periodic table, and all of the subsequent physics and chemistry upon which modern engineering relies. Looking back beyond the scientific revolution of the seventeenth century provides a better understanding of what is unique to engineering and how it has become so closely entwined with science. Until a few centuries ago—relatively recently in historical terms—what science existed had little in common to the practices that we are familiar with today.

Western science has roots going back to classical Greek civilization, to the time of Pythagoras and Plato, with its fascination both with pure mathematics and the place of mathematics in the study of music and astronomy. The Greek scientific tradition continued for centuries, into the Roman era. Alexandria, Egypt, became the center of scientific investigation: the study of geometry culminated in Euclid's systematic exposition, and Ptolemy brought together astronomical observations and theory in a mathematical synthesis that remained unchallenged until the Renaissance. But the science of antiquity was highly theoretical, centering on geometry and observation rather than experiment. Mathematicians, astronomers, and natural philosophers, as scientists were then called, had little interest in studying earthly problems. Had their curiosities been more earthbound, their studies would have been more relevant to the practical arts and to the technological challenges of the time.

Nevertheless, there were exceptions. The great scientist and mathematician Archimedes set forth the mechanical principles governing levers, wedges, screws, pulleys, and other machine elements, as well as hydrostatic principles that remain useful to today's engineers. Later, in the first century BCE, Hero of Alexandria invented ingenious mechanical devices, one of which converted heat from steam to mechanical motion, eighteen hundred years before the invention of the steam engine! But Hero's devices found little practical use, for he built them only to awe and amuse the aristocratic few with his ingenuity. Archimedes, in contrast, became widely known for his engineering acumen in devising sophisticated fortifications, catapults, siege engines, and other military equipment for the defense of his native Syracuse against the invading Romans. But in keeping with the tenor of antiquity, he downplayed his practical accomplishments, placing far greater emphasis on his theoretical work in mathematics and pure science.

Physics, arguably the science most relevant to engineering, became codified in the voluminous writings of Aristotle. Aristotle's approach to natural philosophy gained such authority that it stood virtually unquestioned for nearly two millennia. His followers eschewed experiment and turned the study of nature into purely speculative intellectual inquiry. From intense pondering, they formulated concepts to explain the behavior of natural world. Their conclusions came through formal argument, and they judged the validity of their explanation by its universality and ability to satisfy reason. They thought they could understand nature solely through pure and intense thought, speculation, reflection, meditation, and debate. They searched for first causes and aesthetically appealing reasons for why things behaved the way they did, but they showed little interest in describing in detail with mathematics how nature behaved. Experiments had no place, in their view; for such undertakings would be pedestrian and perhaps even misleading. Moreover, experiments, and the physical labor they might require for building apparatus and obtaining data, were beneath the intellectual and social status of philosophers of nature.

In contrast to the speculations of natural philosophers, craft workers through the ages labored to produce the artifacts that attempted to meet civilization's material needs. For the most part, progress and improvement were at a slow, nearly imperceptible pace. The limited knowledge, tools, and capital available could result only in incremental improvements in a process likened earlier to the slow, Darwinian evolution of plant and animal species. And yet such incremental progress was punctuated on occasion by technological innovation of a more radical sort. The arch, the medieval horse harness, the ship's rudder, waterwheels, windmills, and plows are only a few of the inventions that preceded the rise of modern science, the science that began to take shape during the Renaissance.

That we will ever understand with any certainty the innovative processes that engineers of earlier times followed in bringing about such innovations is unlikely. But some reasonable conjecture may shed light on the general nature of how they engineered these artifacts. The wheel, once again, illuminates some of these issues. To us, the advantages of the spoked wheel over its solid predecessors are clear. Spokes reduced weight and thus increased energy efficiency; their resilience also cushioned rides over rough terrain. But what about the ancient engineers—how did they come up with such an idea? How did they know the benefits that would result, and how did they bring the spoked wheel into existence?

Clearly, they first had to visualize the concept before they could actually build such a wheel. Did the first wheel they built succeed in meeting their expectations? Probably not. More likely, they had to try or test the wheel in some way. And then from what they learned of its shortcomings, they changed or modified it, rebuilt it, and tried it again. The process might be diagramed something like that shown in figure 32. For the most part, tacit knowledge allowed the craftsman or inventor to express the new idea directly as the built artifact. In architecture and ship construction, drawing before building played an important role even in the craft guilds of medieval times, and probably even earlier. And with the revolution in drawing that they brought about, Renaissance engineers advanced the art of setting down their ideas, refining them, and examining alternatives on

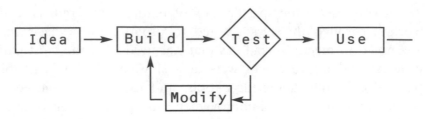

Figure 32. The trial-and-error method. (Diagram by E. E. Lewis.)

paper before they attempted to implement their idea as an artifact. But there was little science in the process, at least not science as we would recognize it today. We would generally characterize the process simply as trial and error.

Where the inventive idea for a spoked wheel came from we can only guess. Nor do we know how the ancient engineers reasoned why their wheels failed—when the spoke broke or the rim collapsed. How did they conspire to salvage the concept—by adding more or thicker spokes, for example—and modify the design? Still, this much seems certain: they must have combined visionary imaginations capable of conceptual leaps with the tacit knowledge acquired from intimate familiarity with the wheelwright's trade.

How did engineering evolve from the intuitive methods of these ancient practitioners to our present-day practices? Scientific methods now permeate the design process: in the generation of new concepts, in the prediction of how the artifact built according to these concepts will perform, in the testing of the built artifact, and in the analysis of its behavior. The integration of science into engineering has a long and complex history that parallels the development of science itself. To gain some insight into the long struggle that brought about the science-engineering synthesis, we must go back to the seventeenth century. For it was then that the scientific revolution marked a radical turning point in the study of nature, a throwing off of the deference to ancient philosophical authority, and a newfound embrace of detailed observations, direct experimentation, and careful measurement. More than any other, historians associate the genius of Galileo Galilei (1564–1642) with these profound changes in scientific methodology. They should also credit him with pioneering developments that were also to lead to revolutionary transformations in engineering practice.

❀   ❀   ❀

Popular accounts of his life most frequently center on Galileo's building of his own telescope and turning it to the sky to make astronomical observations of momentous importance. He observed sunspots, the rocky nature of the moon's surface, and the moons orbiting Jupiter. His discoveries and his advocacy of the Copernican theory—that the earth orbited the sun—led him to publish *Dialogue on the Two Chief Systems of the World* in 1632. At odds with the religious doctrine of the time, his assertions that the same physical laws governed the heavens and the earth landed him in trouble with the church hierarchy. In what is one of history's great humiliations, he was forced to recant his views and live under virtual house arrest thereafter.

But there is another side to Galileo's genius. Like Leonardo before him, Galileo devoted a good part of his career to engineering practice and was closely involved with the technological developments of his time. He took part in a variety of engineering projects in the Italian cities of Florence, Padua, and Venice and played an active role in the large engineering reclamation projects intended to improve the agricultural production of the hinterlands of Venice. He also was a consultant for the building of military fortifications for the northern Italian city of Palmanova. Later, he brought his engineering acumen to bear on the drainage and flood control projects around Florence. He served as superintendent of waters, wrote a report on the river Bisenzio, and successfully patented a horse-powered pump that raised water for irrigation.

Such experiences fed his dissatisfaction with the trial and error methods employed in these engineering projects. He focused on the problem of formulating more rational engineering procedures. An avid interest in structures and machinery comes through in his writings. He taught his students not only pure science but also how to build fortifications and other structures. The discovery of scientific principles and the formulation of engineering procedures are closely intertwined in much of the work for which he is famous. Toward the end of his career he published the book *Two New Sciences*. In addition to setting forth seminal contributions to the scientific method, the book's treatment of strength of materials and of dynamics may in many respects be considered the first textbook in the engineering sciences: the sciences of man-made artifacts.

Galileo's interest in making engineering practice more rational and scientific was not new. Engineers since the early Renaissance and before had been fascinated by mathematics and were strong proponents of demonstrating its power in their practical pursuits. They looked to mathematics to free engineering from its reliance on tradition-bound guild practices. But even in Galileo's time, more than a century later, the promise of mathematics had not been demonstrated in engineering practice. It is important to understand, however, that in the West, mathematics for centuries had been synonymous with geometry. The algebra developed in the Islamic Middle East had not yet been assimilated into European mathematics. The ubiquitous formulas that populate engineering textbooks today were unknown; representing an unknown quantity with x was still in the future, and the introduction of Newton's calculus was still many decades away.

Geometry had great aesthetic appeal, and it played a prominent role in architecture, where proportions and shapes of structures were foremost in designers' minds. Moreover, as practiced long before even the building of Gothic structures, the properties of stone—great strength in compression, but practically none in tension—served to make the geometry of stable building configuration the foremost consideration in masonry design. Geometry was indispensable in planning fortifications as cannons became more powerful and their fire more accurate. Fortified cities had to be laid out geometrically so that defending gunners had clear lines of fire to all approaches; no part of the walled city could offer a blind spot for the attackers to approach. Thus in stone, brick, or concrete buildings, and even more in fortifications, geometrical reasoning was central to the design process.

Geometry by itself, however, was not very helpful in attacking many of the problems faced by Renaissance engineers. For if only geometric principles were required, they should have been able to scale a structure or a machine to a larger size simply by increasing each dimension by the same ratio. If they needed a bridge that was 25 percent longer than one that existed, for example, they should have been able to build the new one simply by increasing each of the dimensions of the present bridge by 25 percent, and everything should have been fine. But it fact it was not; it seemed, rather, that if a larger structure or machine replaced smaller ones with the same proportions, it became weaker; it tended to sag or even break.

Failures of geometric scaling were ubiquitous. The Renaissance fascination with machinery and the ability to design conceptually on paper resulted in a plethora of machines proposed for pumping, hoisting, spinning, drilling, lifting, and other useful tasks. Although many of these existed only as drawings, never to be built, their designers sometimes succeeded in building small scale models that operated as planned. But even if the model worked perfectly, the full-size machine, scaled up by geometric principles, invariably would fail; it would sag, bind, or sometimes break under its own weight. Similar problems occurred in bridges, buildings, ships, and other structures as well.

This problem of scale was the design dilemma that occupied Galileo's attention during his visits to the great shipyards of the Venetian Republic. As they built ships of increased size, the shipwrights found that the scaffolding required to support the hull had to be disproportionately larger. Why was this the case? Nothing in the geometric arsenal of the engineers of the late Renaissance explained it. Galileo understood that stone and wood behaved differently even for columns, beams, or shafts made in exactly the same size and shape. Geometric rules did not account for these differences, since geometrically they were identical. Leonardo da Vinci had struggled with this problem earlier and in fact had performed some testing of the breaking strengths of materials. But Leonardo's work was lost for centuries, and the relationships between strength, size, and geometric proportions remained confused at best.

Galileo's genius generalized the puzzles of shipbuilding to size limitations found in nature. Just as there appeared to be limits on the sizes of ships and the scaffolding needed to support them during construction, so were there limitations on the sizes of animals and of plants. Why were there no animals larger than elephants, and why did there seem to be a natural limitation on how tall trees would grow? The long, thin wiry frames of insects supported them quite adequately. Men and horses, however, had bulkier bone structures, bones that were thicker in proportion to their length. And in relation to the size of its body, the bones of an elephant were most massive of all.

To understand the failure of geometric scaling Galileo used the theory of the lever, inherited from Archimedes, to analyze a simple beam, representing it as shown in figure 33—a picture that was to become famous in the history of science. Most important, he found the beam's strength to be proportional to its width and proportional to the square of its height, but inversely proportional to its length. Thus beams' cross-sections, even today, are always tall and narrow to maximize strength and minimize material. A beam, however, must support its own weight, and its weight is proportional to its volume (height × width × length). Thus as the beam becomes longer and weaker, its weight becomes larger. With increasing length, the weight that the beam can support becomes smaller and smaller. At some length the beam is able to support no added weight at all and breaks under its own weight. Although Galileo's analysis contained some erroneous details, which were later corrected, his most important conclusions stand today.

One of the more celebrated outcomes of Galileo's studies is the so-called three-two law of structures. The weight of a column, whether it is a tree trunk or a pillar from a Greek temple, increases with volume (three dimensions), while its strength increases with cross-sectional area (two dimensions). Thus if you make a column large enough by increasing all the dimensions by the same ratio, it will eventually be crushed at its base from its own weight. For a beam, the analysis is subtler, but a similar result holds: if you continue to increase all of the dimensions (including length) keeping the shape the same, the beam will eventually break from its own

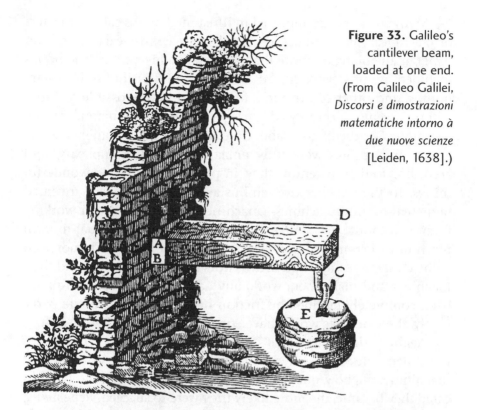

Figure 33. Galileo's cantilever beam, loaded at one end. (From Galileo Galilei, *Discorsi e dimostrazioni matematiche intorno à due nuove scienze* [Leiden, 1638].)

weight. Parenthetically, while the column fails from crushing (in compression) the beam fails from stretching (in tension). Thus, as mentioned in an earlier chapter, since stone has great strength in compression but hardly any in tension, stone columns are a familiar sight, but you will never see a stone beam.

Galileo's success in relating geometry to the strength of materials provided the beginnings of the engineering science of structures, and of machinery as well. But to formulate a useful theory of machines, he had to go further and unscramble some widespread misconceptions of the time. In the sixteenth century, the belief was pervasive that machines and other material artifacts had spirits and that these could be tricked to give up their powers. Frequently, an inventor's object was to trick or confound nature into giving up something for nothing, and the more elaborate the machine, the more likely it was to succeed.

Many such devices appeared in illustrated books called "theaters of machines," as noted earlier. But these machines existed only on paper, many being no more than speculative inventions. The books included pumps with curved cylinders and piston rods, rotary pumps, and other contrivances that would be impossible to build without twentieth-century advances in materials and manufacturing techniques. Exceedingly elaborate combinations of pulleys, gears, levers, and springs were truly amazing in their complexity, and according to their inventors they, in principle, would do wonderful things. In their exuberance, enthusiasts claimed to have invented perpetual-motion machines—machines that would go on working forever without any source of outside power. Perpetual-motion machines, of course, cannot exist, and their devices would not even come close. In reality, if these inventors had been able to build their machines and make them work, any advantage gained would have been completely nullified by friction between their elaborate parts. Likely, the machines would have come quickly to a grinding halt.

Against this backdrop, Galileo launched his studies of machinery. He cleared away many of the mystical notions that machines somehow outsmarted nature and that the more complicated they became, the more likely they were to succeed. Repeatedly, he attacked the confused thinking that machines could provide something for nothing or create forces or power out of nowhere.

Galileo sought to explain mathematically the advantage of using machines. Reaching back to Archimedes' studies, he looked at the basic mechanical elements from which the machinery was composed and made a number of important contributions to the scientific explanation of their operations. He analyzed more complicated machines in terms of their basic components and motions. He demonstrated that machines functioned to change the rate at which work is done (such as gearing that allows a slow-turning waterwheel to drive a faster-rotating sawblade or grindstone). They also provided transference of effort from one place to another (such as a pump moving water up from a well), but they didn't get something for nothing.

The science of engineering mechanics originated in Galileo's work. Through Galileo we can also trace roots of a more general problem-solving technique now inculcated in all engineering students, regardless of their field of specialization. Like Galileo, present-day engineers use scientific principles expressed in mathematics to analyze complex devices and systems in terms of the properties of their elementary components and the interactions between them. Civil engineers analyze structures in terms of the beams, columns, arches, and other forms from which they are built. Mechanical engineers view machinery in terms of the levers, springs, gears, valves, bearings, and other components from which it is made. Electrical engineers derive the behavior of complex circuitry in terms of the resistors, capacitors, inductors, and transistors from which it is composed.

We can attribute to Galileo more than an objective and mathematical approach to the analysis of man-made artifacts. The engineering considerations of efficiency and cost of operation played a central role in his studies. He ranked the greatest advantage of machines as their ability to harness existing sources of power, wind, falling water, and animal muscle and apply them economically to useful tasks. He introduced economics into machine design by comparing waterwheels and wheels turned by horses' power and assessing their relative costs of labor and maintenance. He was interested in more than academic mechanics; he was a forerunner in bringing together physics and mathematics with economics, an amalgamation that is present throughout engineering today.[1]

Closely related to Galileo's elucidation of machine principles was his work on dynamics: the science of motion. We often envision the apocryphal picture of Galileo dropping different sized objects from the Tower of Pisa. As the story goes, he observed the contradiction of the Aristotelian teachings, noting that in fact the heavier object did not fall faster than the lighter; they both reached the ground at the same time. In reality, he devised many more exacting experiments in which he slowed the falling motion so that he could measure it more accurately. To accomplish this he meticulously constructed an inclined plane with carefully formed grooves to minimize friction and other unwanted side effects. He then rolled brass balls down along the grooves and measured the time it took them to travel over marked distances. He devised a water clock to measure time; the time elapsed was proportional to the volume of water collected by his apparatus.

Combining his carefully executed measurements with mathematical analysis, Galileo formulated rules of motion. His studies did more than define the methodology for performing and analyzing experiments; they also became basic building blocks for profound advances in both the pure sciences and in developing principles for engineering analysis. The rules of motion that derived from his experiments formed the basis for the future understanding of motions ranging from the orbits of planets to the trajectories of cannonballs.

Galileo's studies in pure science became one of the cornerstones that led to a revolution in astronomy. A second came from his contemporary Johannes Kepler. After careful study of the massive but exacting observations accumulated by the Danish astronomer Tycho Brahe before the telescope's invention, Kepler showed that the data obeyed three rules. The first stated that the orbits of the planets were not the perfect circles of time-honored natural philosophy, but ellipses. The second and third rules related the size and shape of the ellipses to orbital velocity and period of time elapsed during an orbit. Then, in 1687, less than a century later, the *Principia*, the monumental work of Sir Isaac Newton, appeared.

In the *Principia* Newton stated his famous three laws of mechanics and put forth the universal law of gravitation. In addition to demonstrating that his gravitational law explained both Galileo's rules of motion and Kepler's elliptical properties of orbits, Newton established the system for analyzing motion in terms of the forces that cause them. His system has stood as the foundation of physics for over three hundred years, requiring elaboration only in the twentieth century to allow for Einstein's relativity theory. Thus we can trace back to Galileo a sequence of events that in roughly fifty years brought astronomy from a state of philosophical speculation to an objective mathematical science.

Galileo's combination of experiment and mathematics also had more earthly effects. In what is perhaps the most elegant piece of applied mathematics found in *Two New Sciences*, he established that in the absence of air resistance, the flight of a projectile, such as a cannonball, takes the form of a parabola. Moreover, a ball leaving the cannon barrel with a given speed will have the greatest range if it is launched at an angle of forty-five degrees to the horizon. Clearly, this analysis was relevant to gunnery problems of the time. However, its successful application to solving practical problems of military engineering was much slower in coming than was the impact of Galileo's work on astronomy. The reasons for these delays are instructive, for they provide us with perspective on some of the complications involved in applying scientific principles to predicting the behavior of engineered artifacts. They illuminate why it proved more difficult to predict the range of a cannon than to explain the motions of planets.

The arrival of gunpowder from China in the fourteenth century was a momentous event of European history. Europeans rapidly learned to craft workable cannons, first from brass and then from iron, adapting, ironically, the metal-casting process used to make church bells. Soon the cannons fired projectiles with destructive power vastly superior to that of the stones hurled by traditional catapults. Warfare was revolutionized: medieval castles and fortified cities became indefensible; for by time of the early Renaissance, cannon fire could reduce their tall but thin masonry walls to rubble in hours. Advances in gunnery necessitated construction of new fortifications. And from Leonardo da Vinci to Galileo, prominent engineers of the Renaissance engaged in designing and building fortifications capable of both absorbing the punishment of the increasingly powerful artillery fire and providing defensive gunnery with clear lines of sight to rain havoc on approaching artillerymen. Likewise, gunnery transformed warfare at sea. Galleys, which had been rowed into battle since early Greek civilization, quickly became obsolete. The sailing men-of-war that replaced them were capable of delivering broadside salvos of overpowering cannon fire.

Gunnery, however, developed as a very empirical art. For centuries, gunners loaded powder and aimed their cannons by rule of thumb. What little theory there was still relied on an Aristotelian theory of flight even after Galileo had discredited it. Long after Newton's celestial mechanics explained the planets' orbits, the engineering problems associated with predicting where a cannonball would land remained intractable. They baffled even the best scientific minds of the time.

Careful comparison shows that prediction of a cannonball's range encounters difficulties not found in celestial mechanics. The telescopes, clocks, and other apparatus that seventeenth-century astronomers used in making their observations left much to be desired. But to their advantage, the only errors introduced into their observations of planetary orbits were those caused by the imperfections in their instruments and shortcomings in their methods of measurement; the orbits themselves remained unchanged. Thus they could repeat their observations, correcting them for earlier errors and inaccuracies. The mathematical precision of the heavenly clockwork, however, was entirely different from the problems faced on the ground by military engineers as they struggled to employ scientific approaches to predict where their cannonballs would land.

The first difficulty was that no two cannons were exactly alike. Craft workers cast barrels in bronze or iron and bored them using techniques that are crude and imprecise by today's standards. Even small variations in diameter and differences in roughness from barrel to barrel gave rise to much larger differences in muzzle velocity and therefore in the distance that the cannonball traveled. Thus the imprecise casting and boring techniques meant that experimental results obtained on one cannon would be significantly in error for another. Such errors occurred even though the barrels were of the same design; they were even crafted according to the same engineering drawings.

Other imprecisions of craft practices added to these problems. The diameters and weights of the cannonballs varied considerably, as did the uniformity of the gunpowder. Thus the range varied not only from cannon to cannon, but from firing to firing from a single cannon. Further complications resulted from lack of durability. With repeated firing, barrel wear introduced yet another source of variability into the trajectory.

These complications, none of which were present in the study of planetary orbits, typified those encountered in attempts to apply scientific principles to the engineered artifacts of the preindustrial age. They indicate why successful analysis of structures and machinery was delayed until long after Galileo, long after Newton, and well into the industrial revolution of the eighteenth century. Scientific analysis requires that reproducible experiments be performed. In pure science this means that the experimental apparatus and the surrounding environment must remain constant from trial to trial. For engineering analysis to be successfully applied, however, the artifact under study must be reproducible. The capability must exist to replicate it with sufficient precision and durability that variability in dimensions, material properties, surface roughness, wear, and the like do not obliterate the phenomena under study. Such capabilities came into being only as more precise production by machine replaced traditional handcrafting methods.

In gunnery, the complications did not end when the cannonball exited the barrel. For unlike the vacuum in which the planets move, the cannonball's flight was within the earth's atmosphere and thus encountered air resistance. But as Galileo clearly stated, his parabolic trajectory omitted the effects of air resistance. That omission confined the theory's validity to predicting the ranges of only the slowest-moving mortar shells, where air resistance had little effect. For most trajectories air resistance was important, and unfortunately, it varied with temperature and humidity. Thus even if balls could be fired from a cannon barrel with a precisely controlled velocity, their range would vary with the time of day, with the season, and with local weather conditions.

These complications made application of mechanics or other developing sciences to the artillery problem very difficult. As precision manufacture improved, variability in the cannon, the ball, and the gunpowder became less of a problem. The challenge to understanding then centered about the muzzle velocity. The internal ballistics, or what went on inside the cannon—the detonation of the gunpowder and the acceleration of the ball down the barrel—determined muzzle velocity. Muzzle velocity, in turn, was the cannon's dominant influence on external ballistics, the flight of the ball through the atmosphere. However, muzzle velocity was not measurable with clocks, rulers, scales, or other scientific instruments of the time. As in many other engineering challenges, rapid progress had to wait on the invention of a critical piece of measurement apparatus, in this case apparatus to measure muzzle velocity.

The salient advance in experimental apparatus for artillery pieces was Benjamin Robins's invention in 1742 of the ballistic pendulum. This apparatus, combined with a great deal of experimentation and mathematical analysis, finally allowed the engineering science of ballistics to make rapid progress.[2] With the ballistic pendulum, shown in figure 34, muzzle velocity was measurable. The cannonball would hit the plate at a right angle and swing the pendulum arm. From the pendulum's length and its angle of swing, and from the weight of the pendulum and the ball, Robins could make accurate estimates of the projectile's speed as it hit the plate. As a result, artillery engineers could separate their studies into interior and exterior ballistics, into what went on inside the cannon and what happened during free flight.

Robins used his ballistic pendulum in extensive experiments and combined his experimental results with those of others. With an understanding of the theoretical work of Newton, Boyle, Bernoulli, and other scientists of the day, he worked out a theory of expanding gases. From this he was able to calculate the muzzle speed in terms of the weight of powder, the length and diameter of the gun barrel, and a number of other parameters. This work on interior ballistics thus brought scientific analysis to the design of artillery pieces. For now science had been added to the diagram shown in figure 32. If engineers designed a new cannon, they could apply Robins's theory to their drawings and estimate the muzzle velocity before they built the prototype. They now had mathematical design rules to guide their work. Likewise, once they had the cannon built they could use the ballistic pendulum to conduct tests on the prototype to confirm that the design

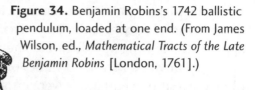

Figure 34. Benjamin Robins's 1742 ballistic pendulum, loaded at one end. (From James Wilson, ed., *Mathematical Tracts of the Late Benjamin Robins* [London, 1761].)

met their muzzle velocity criteria. If it didn't, the test data focused attention on the limitations of the design rules; the data guided the redesign of the cannon and the design rule revision.

If anything, external ballistics problems were even more daunting. The ballistic pendulum could determine the muzzle velocity, and with that known, Galileo's parabolic theory predicted the range. Nonetheless, since Galileo did not include air resistance, very large discrepancies resulted: high-speed projectile ranges were as little as one-fifth of those predicted by the parabolic theory.

But here, too, engineers successfully employed the pendulum apparatus. By measuring the projectile's velocity at two different distances from the muzzle, they determined how much speed it had lost, and from that they estimated the magnitude of the air resistance. Firing the cannon repeatedly under carefully recorded atmospheric temperatures and humidity, they could then correct the air resistance value for changes in atmospheric conditions. A second piece of apparatus, the whirling arm, also invented by Robins, allowed further study of air resistance at lower velocities, where use of the ballistic pendulum was problematic.

Previously, Sir Isaac Newton himself had derived equations for projectile flight with air resistance included. But without values for the muzzle velocity and the air resistance magnitude, he could not apply them to range predictions. The new experimental apparatus, however, provided measured values of the necessary quantities. Even then, the resulting equations were intractable. Earlier, Newton, using the calculus he had invented, and others had wrestled without success with these complex, nonlinear equations. It was preeminent eighteenth-century mathematician Leonhard Euler who invented a laborious but successful approach for obtaining numerical solutions. Euler's method allowed engineers to combine theory with experiment to obtain estimates of a cannonball's range.

Other successes arose from the extended effort to apply scientific approaches to the problems of cannon fire. Experimentation by Robins and his successors clarified the effects of artillery shell shape and other properties on air resistance. Probably the most famous of these was the "Robins effect," the discovery that rifling of gun barrels resulted in more accurate fire. Spiral grooves along a gun's barrel, called rifling, impart a spin to the bullet that increases the accuracy of its trajectory.

Robins's experimental observations also disclosed a very sharp increase in air resistance at velocities of roughly eleven hundred feet per second. He observed that this was approximately the speed of sound in air and suggested further experimental observation of the effects. He was thus an early observer of the so-called sound barrier. He provides one of the earlier instances in which study of engineering problems through scientific methods uncovered curious natural phenomena and led subsequently to fundamental advances in our understanding of nature. Such interplay of science and engineering has become ubiquitous in more recent times. Likewise, centuries later, with the invention of the digital computer, Euler's method and others stemming from it still serve as a basis for obtaining numerical solutions to complex equations in both engineering and science.

※　　※　　※

The contrast between Newton's explanation of planetary orbits and the difficulties that delayed the ability of artillery engineers to predict cannonball trajectories indicates some of the barriers in using scientific methods to gain understanding of the behavior of engineered artifacts. Nasty problems plagued engineers in applying the accumulation of scientific theory and methods to machines or structures: they included the perplexingly difficult-to-analyze geometrical configurations; the prominent effects of friction and wear; the imperfect and poorly understood properties of the materials of construction; and the imprecision of craft production methods. Nevertheless, what Robins and Euler accomplished in military engineering was also indicative of advances made in what became known as civil—that is, nonmilitary—engineering. Across a range of endeavors, the eighteenth century saw an increasing pace of success in applying mathematical analysis and experimental method to the prediction of the strength of structures and the efficiency of engines.

The difference between the careful experimental emphasis of Robins, who was British, and the mathematical approach of Euler, who was Swiss, symbolizes in some sense the differing emphasis taken in Britain and on the Continent in integrating scientific method into engineering. Continental efforts, particularly in France, stressed the mathematical theory that underlies engineering applications, viewing experimentation as a way to verify their theoretical work. British engineers were of a more empirical bent, working from testing and experiment toward the rules, formulae, and theory that would prove invaluable in translating their ideas to artifacts.

A number of factors contributed to French leadership in rationalizing the practice of engineering. French highways, bridges, and other infrastructure of the eighteenth century were the most advanced in Europe. The bureaucracy of the centralized government provided a core of scientifically trained engineers to improve that infrastructure and to enhance the nation's economic and military power. The French published the first engineering science texts and established the formal education of engineers based on scientific principles. The first professional engineering college, the École des Ponts et Chaussées (School of Bridges and Roads) was established in 1747. Then, following the revolution, Gaspard Monge, who also invented descriptive geometry, founded the École Polytechnique in 1794. He instituted a three-year curriculum in which the first two years consisted of basic sciences and mathematics. The resulting professionalization of engineering drew it away from the apprenticeship form of craft training and directed it toward the applied sciences. The École Polytechnique served as a model that spread to other countries, most notably Germany and the United States; America's first engineering school, the US Military Academy, established in 1802, was patterned after it.

With state support, the spirit of the Enlightenment, and their flair for abstraction, Continental mathematician-engineers derived theories to explain the behavior of the buildings, bridges, dams, and viaducts that had been built by trial and error for a millennium or more. They applied increasingly sophisticated mathematical analysis to understand the bending strength of beams, the collapse of columns, the pressure that soil placed on retaining walls, and the thrust that arched bridges exerted on their abutments. The fundamentals that they established gave them enduring recognition, witnessed by the frequent reference made by today's engineering textbooks to the pioneering mathematics of Jakob Bernoulli, Charles-Augustin de Coulomb, Leonhard Euler, Jean-Baptiste-Joseph Fourier, Joseph-Louis Lagrange, Pierre-Simon de Laplace, Claude-Louis Navier, and Siméon-Denis Poisson.

The pinnacle of French achievement was mathematical analysis, and the results had lasting effect. But often long delays ensued before the abstract theoretical work, cast in mathematics that few practicing engineers understood, could be refined, verified, and translated into design rules that were readily applicable. Extensive experimentation to quantify the properties of traditional building materials had to complement the theoretical work. Thus, engineers devised apparatus to test the crushing strengths of commonly used stone, brick, and mortar. They devised materials tests to understand what loads different types of wood could withstand without breaking and the degree to which they would sag. They predicted the strength of pipes, relating bursting pressure to diameter and thickness. But even then gaps remained between mathematical formulations and practical application. For the mathematics of the engineering elite was abstract, and practitioners were practical men, often skeptical of abstraction.

In England, too, the beginnings of more systematic engineering methods emerged in the eighteenth century. But there the situation was quite different. In the decades leading up to the industrial revolution, most leading engineers were in private practice. And they were empirical men, trained through apprenticeship rather than formally educated, and thus suspicious of highly theoretical analysis. The problems that most occupied their efforts also differed from those on the Continent, for the gathering momentum of the industrial revolution brought acute energy shortages to England.

As the eighteenth century progressed, the rivers and streams of Britain became overtaxed. With too many waterwheels of ever-increasing size, suitable sites for new factories were few and far between. Power from wind was even more limited. The output of windmills, traditionally used for grinding grain and pumping water, varied too much with the weather to provide the insatiable power demands of the increasing numbers of manufacturing facilities. And power from steam was still in its infancy.

Gaining maximum efficiency from waterwheels and being able to predict how much power a newly constructed wheel would deliver were issues occupying many an engineer's mind. But knowledge was minimal, and controversies raged over even seemingly basic questions, such as whether the overshot or undershot waterwheel was more efficient, and what their efficiencies were. The problem gained the attention of British engineer John Smeaton, who was to become famous during his lifetime for his innovative design of the Eddystone Lighthouse; for his engineering of many bridges, canals, dams, and mill works; and for his effort to establish civil engineering as a profession distinct from military engineering. But the experimental studies that he undertook to investigate the efficiency of waterwheels, and later of windmills and steam engines, were to have an even more lasting impact. They were important for improving energy production in England, but even more so for the logical basis that they brought to the process of engineering design.

Smeaton's extensive experience with water-powered mill works caused him to puzzle over the question of waterwheel efficiency. He was aware of Antoine Parent's theoretical work in France that predicted the efficiency of the undershot wheel to be 14.8 percent. But he thought that this number was far too small and that the theoretical analysis must be deficient. In the 1750s Smeaton initiated a series of carefully designed experiments to determine whether the undershot or overshot wheel was more efficient. He sought to understand how the wheels' performance and efficiency could be predicted from the flow of the stream, the size of the wheel, its speed of rotation, and other related properties. His task was compounded by the confusion in conceptual understanding at the time. Definitions of work, power, and energy as understood by today's engineers had not been established, and the central theoretical underpinning, the principle of conservation of energy, was not fully elucidated for nearly one hundred years after his experiments.

Nonetheless, he proceeded by building apparatus ideally suited to his task. It allowed him to use scale models to predict the behavior of much larger wheels, and moreover he could adapt his setup to test either overshot or undershot wheels. His model wheels were about a meter in diameter, as pictured for an undershot wheel in figure 35, with the water supply coming from an adjoining tank. After passing by the wheel, the water was recycled by a hand pump back to the tank. During a test, he maintained a constant flow rate past the wheel by pumping to keep the tank filled to a fixed level. In successive tests, he increased the flow step by step by increasing the height of the tank. He measured the power of the water at the wheel and the performance of the wheel (the "effect," as it was then called) for various flow rates. To perform these measurements he designed the pulley arrangement, shown in figure 35, that enabled him to record the height through which the wheel raised a standard weight in one minute.

Experimenting with various weights, flows, and heights, he established that the overshot wheel was more than twice as efficient as the undershot. He also deduced a range of optimum values for how fast an undershot wheel should rotate relative to the stream velocity for maximum efficiency. From his experiments, he estab-

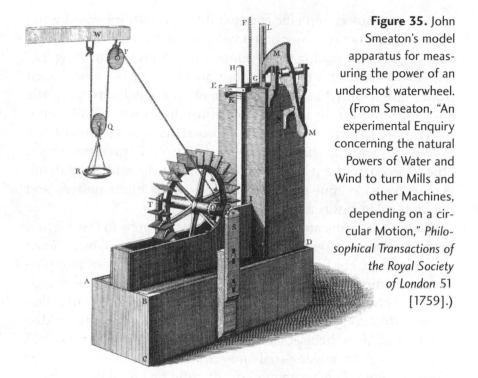

Figure 35. John Smeaton's model apparatus for measuring the power of an undershot waterwheel. (From Smeaton, "An experimental Enquiry concerning the natural Powers of Water and Wind to turn Mills and other Machines, depending on a circular Motion," *Philosophical Transactions of the Royal Society of London* 51 [1759].)

lished the efficiency of the overshot wheel to be approximately two-thirds, and the undershot wheel to be about one-third, much higher than Parent's theoretical prediction. Smeaton's visual observations also were important, for he observed that not only the losses resulting from friction between the wheel's moving parts but also any spray or turbulence that the wheel caused robbed it of efficiency. Thus, to the extent that turbulence and friction could be reduced, higher efficiencies might be achieved. Later, in 1767, Jean-Charles de Borda's theoretical analysis confirmed that an idealized undershot wheel—one without friction or turbulence losses—had a maximum efficiency of 50 percent, achieved when the wheel rim velocity was one-half that of the impacting water. Likewise, a similarly idealized overshot wheel could approach 100 percent efficiency when revolving very slowly relative to the stream speed.

Smeaton's experimental conception, apparatus, and insightful analysis were precedent breaking. He successfully used models to predict how power and efficiency would scale with increased water flow, wheel diameter, rotation rate, and other factors. Far from simple geometric scaling, he developed procedures for treating friction losses and for separating those arising from the wheel itself from those incurred as a result of the experimental apparatus. His later apparatus for the investigation of windmills was equally innovative. The rotating-arm apparatus depicted in figure 36 allowed him to create an artificial wind of known velocity. With this setup he was able to ascertain the relationship of wind velocity and blade diameter to power output and also to test various blade designs and determine which was most efficient.

In the 1760s Smeaton's engineering practice grew to involve him in projects requiring the design and construction of Newcomen steam engines. For half a century, these engines had been pumping water by converting the heat of burning coal to mechanical energy. Nevertheless, the design of these engines still came from intuition gained from long experience. Exceedingly little scientific understanding of their behavior existed. At first, Smeaton approached engine design by making trial-and-error modifications to improve efficiency. But following engine construction he found the perform-

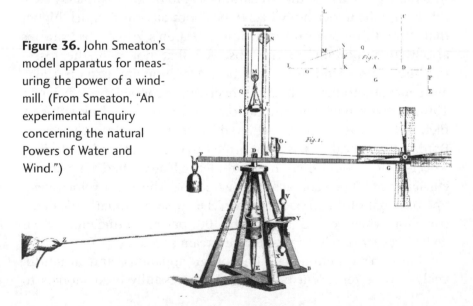

Figure 36. John Smeaton's model apparatus for measuring the power of a windmill. (From Smeaton, "An experimental Enquiry concerning the natural Powers of Water and Wind.")

ance disappointing compared to what he had hoped for. As a result of this setback, he set about to gain a quantitative understanding of the operation of Newcomen engines. For only that understanding would allow him to predict and optimize performance before construction began. His approach followed the same disciplined engineering paradigm that had worked so well in his investigations of wind and waterpower.

He proceeded to build a small experimental engine of a little more than one horsepower. Over a period of two years, he used it to perform more than 130 experiments. He measured temperature and pressure in the cylinder, barometric pressure, and the amount of water evaporated per bushel of coal burned, as well as several other factors. After adjusting the engine to good working order, he would alter one particular factor and then measure the change in performance. He learned to synchronize the timing for valve opening and closing and to optimize the loading on the pistons. He tested different forms of injection nozzles and utilized feed water heating and piston insulation to further increase efficiency, and he studied the relative heating capabilities of a variety of types of coal.

Although no one factor dominated Smeaton's improvements, taken together they led to a major advance in engine design. His experimentation enabled him to derive rules for the optimal engine proportions and characteristics that were applicable over a wide range of sizes. Included were specifications for the stroke length and rate, the rate of coal consumption, the quantity of water injected, and so on. For each size of engine, he predicted the power and efficiency to be expected.

Smeaton combined his experimental work with a systematic study of existing Newcomen engines, making careful observations of some fifteen of them. For this work, as well as that on his own engine, he expressed his results in terms of the engine power, then defined as the "great product," and its efficiency, referred to as the "effect." He thus introduced correct quantities for the measures of performance. Watt later redefined these quantities as horsepower and duty, but the concepts remained unchanged. Smeaton determined the rate at which an engine's efficiency increased with its size and power. His observations also characterized efficiency losses resulting from faulty pumps, poorly bored cylinders, and badly designed boilers.

Smeaton's experimentation, observation, and formulation of design rules and criteria paid off in the engine he designed for installation at Long Benton. When put into operation in May 1774, the engine produced the forty horsepower for which he designed it and came very close to his demanding efficiency expectations. It was 25 percent more efficient than the most efficient engine previously built and 40 percent more efficient than the typical good performance of the Newcomen engines of the period. The Long Benton engine established Smeaton as a leading engine builder of his time. Even more important, it served as further verification of the rational methodology that he brought to engineering experimentation and design. Without the benefit of a scientific breakthrough or an ingenious new invention, he demonstrated a systematic approach toward optimizing the performance of existing technology and toward predicting the performance and cost of new machines before they were built.

Smeaton's name is not nearly as well known to the public as that of James Watt, for he made no invention of the magnitude of Watt's separate condenser for the steam engine. The separate condenser, as well as Watt's ancillary innovations, so enhanced energy production with steam that it quickly supplanted the less efficient Newcomen engine and eventually led to steamships, railroads, and a host of other new technologies. Nevertheless, Smeaton's work had a lasting impact on the way engineers practice their profession. The preponderance of engineering endeavors do not center on the introduction of radical new inventions of the historic proportions of the Watt engine. Rather, they are focused on choosing, improving, and optimizing designs by using existing technologies and on adapting them to particular applications and clients' needs. And here Smeaton's procedure, sometimes referred to as the "method of parameters," is fundamental. Analysis, computation, and experimentation using models, mock-ups, and sometimes full-size equipment underlie the paradigm of present-day engineering. These tools enable performance prediction before construction, and, once an artifact is constructed, systematic testing to verify that performance goals have been met.

The increasing ability to derive design rules for predicting behavior before construction and to perform objective verification experiments allowed engineers to be more inventive. It allowed them to venture further from traditional craft practices, which confined them to slow evolutionary improvements, and to strive for more innovative solutions to technical problems. But if integration of scientific methods allowed engineers to venture further from tradition-steeped craft methods, revolutionary technological changes that accompanied the industrial revolution demanded it. With time, the Watt steam engine surpassed wind, water, and animal muscle in providing power for processes, from grinding grain to propelling ships. Coal replaced wood, and coke replaced charcoal as the fuels for heating industrial processes and smelting iron. And as the production of iron in large quantities became more economical, it replaced wood as the primary material for constructing machines and structures. New engineering paradigms were absolutely required to capitalize on these advances, for the existing crafts had little experience to relate to the challenges that these new technologies presented.

Builders of buildings and even more of bridges moved away from the traditional forms of wood and masonry construction to take advantage of the strength and economy of iron. New structural concepts were necessary to gain fire protection by replacing wooden structures with multistory factory buildings made of iron. What is more, the ironwork of new, longer-spanned bridges contrasted dramatically with the stone-arched construction dating from antiquity. Failures, sometimes disastrous collapses, took place all too frequently as engineers struggled to understand the characteristics of iron construction. Overcoming such problems required structural engineers to resort increasingly to careful analysis and experimentation. Safety and economy required them to predict strength before construction and to devise proof and verification tests to confirm their predictions.

The example of famed British bridge builder Thomas Telford is instructive, not so much for the bridges themselves but for the considerable contributions he made toward further systematizing the engineering design and construction of large structures. Telford estimated the tensile forces in cables. He tested hundreds of bolts, wrought-iron bars, and chains to destruction, to find the load under which they would break. He initiated the safety factor concept and designed iron cable so that it was stressed to no more that one-third of its breaking strength. He used mock-ups to test structural configurations before a bridge was built. He used the arrangement shown in figure 37, for example, to test spans as long as 900 feet. He also built models of bridges. On his own property he constructed a model 105 feet long and ran loaded carriages over it to examine its sag and strength. Even more ambitious, he built a quarter-scale model with a span of 142 feet to resolve design questions relating to what was to be his greatest accomplishment, his suspension bridge over the Menai Straits in Wales, completed in 1826.

Robins, Smeaton, Watt, and Telford from the more empirical English school, the French engineer-mathematicians of a more theoretical bent, and numerous others from the era of the industrial revolution all contributed to the methodology that brought engineering closer to the science-based discipline that it is today. The radical changes that were required to capitalize on the potentials of steam power and iron construction came to fruition in industrial mechanization, railroads, and other dramatic technological advances of the nineteenth century. Among the foremost of these was the engineering of the transoceanic steamship.

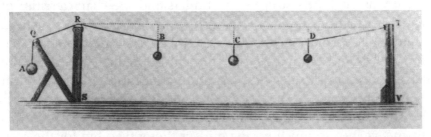

Figure 37. Thomas Telford's test apparatus for suspension bridge cables. (From John Rickman, ed., *Life of Thomas Telford, Civil Engineer* [London: Payne and Foss, 1838].)

Since antiquity, shipbuilding, as noted earlier, had been among the most complex and highly organized of activities in the creation of technology. The procedures and methods evolved slowly with the knowledge and skills passed down from generation to generation through the shipwrights' guilds, punctuated only rarely by inventive changes in hull, rigging, or rudder design. The complexity of ship structures and the coordinated efforts of large numbers of workers required in construction brought the early codification of designs in drawings and the "laying-out" process illustrated earlier. The lines of the hull and the efficacy of the sail plan and rigging were determining factors in the speeds that a ship could achieve, its maneuverability, and its seaworthiness in rough weather.

Variability in wind and weather had larger effects than nominal design improvements in determining how long sailing ships took to reach their destinations. But provided that they withstood the storms, that food and water didn't run out, and that navigation didn't fail, the wind would eventually get them to their ports of call. A ship powered only by steam, however, faced a new and different set of circumstances. Its travel time depended much less on wind or weather. If it ran out of fuel, however, or if its engines failed, it would be unable to reach its destination at all.

The first half of the nineteenth century brought a double whammy to the practice of shipbuilding. The change in power from wind to steam and in hull construction from wood to iron brought more radical changes in design and construction than had occurred for centuries before. In this transformation, which also resulted in the replacement of traditional craft methods with those that more closely resemble the science-based engineering of today, Isambard Kingdom Brunel played a pivotal role. Already acclaimed for his accomplishments in constructing the Great Western railroad across England, he turned his attention to the sea with the goal of extending the English transport network to North America and beyond. To this end, he engineered three oceangoing steamships during the middle decades of the nineteenth century, each of monumental proportions, and each showing the growing importance of scientific considerations in engineering practice.

The antecedents of Brunel's efforts began to appear in the closing decades of the eighteenth century. In the rivers and coastal waterways of Europe and North America, experimental vessels began using steam as a power source. Ironmasters began building primitive hulls of iron, even though their understanding of the material's properties was minimal. But the efforts to use steam propulsion and iron hulls on protected waters notwithstanding, it was common belief among the seafaring community that no ship could be built to cross the Atlantic under the power of steam alone. Vessels the size of oceangoing sailing ships, they thought, could not carry enough coal in their hulls to cross the Atlantic. They believed this to be true even if they were to use the most efficient steam engines available. Moreover, seafarers reasoned that even if the size of the hull capacity could be doubled, for example, the power of the engines needed to drive the ship at the same speed must also be doubled. And if that were true, the volume of coal that the ship must carry would also double. Thus no matter how large a ship was built, its coal-carrying capacity would be insufficient for a successful transoceanic voyage.

Indeed, the first oceangoing steamships were no larger than the sailing ships of the day and could carry enough coal to utilize their engines only for small fractions of their voyages. The engines were valuable for maneuverability and for making progress when winds were calm, but sails were their primary means of propulsion. Brunel's goals for steam were more ambitious: his ships were to be capable of crossing the ocean under continuous power of steam. Yes, they also would carry sails, but the sails would be needed only to stabilize the hull in heavy weather and as an emergency backup should the engines fail.

Brunel's path to creating a steamship capable of transoceanic crossings was to build much larger hulls. He was confident that this was an effective approach because he saw the error of the argument that had dissuaded tradition-bound mariners from building oceangoing steamships. Their argument contained a fallacy of geometric scaling similar to that encountered by Galileo centuries before. But Brunel, like Galileo, understood that he could not simply double all the dimensions of a structure and assume that its other characteristics would increase in direct proportion.

He reasoned that the coal- and cargo-carrying capacity was proportional to the hull's volume. From following scientific developments in fluid mechanics, however, he understood that flow resistance was proportional not to volume but to the surface area of the hull below the waterline. Moreover, it was flow resistance that determined the engine power required to obtain an adequate cruising speed. If he doubled the length and all the other dimensions of the hull, thus not changing its shape, its surface area and thus its flow resistance would increase fourfold. Thus its engine power must also be increased fourfold. But doubling the hull's dimensions would increase its volume—and therefore its capacity to carry fuel—eightfold! Thus if he could build large enough steamships, they could carry enough coal for transoceanic voyages, with capacity to spare for cargo and passengers.

He put this theory to practice in his first ship, the *Great Western*. In its design, Brunel analyzed a number of interacting factors to match hull size, fuel capacity, and power requirements before construction began. After sizing the wooden hull, he estimated its water resistance and from that the power required to propel it at a sufficient speed for attractive transoceanic crossing times. He then procured engines guaranteed to produce that level of power and efficiency needed to drive the vessel's paddlewheels. For from power, efficiency, and voyage duration he was able to estimate the weight of coal that the ship must carry to reach its destination. Like all design, this was an iterative process, fraught with uncertainties. If the engines and the coal required more space than the hull could carry, he had to reassess its size and repeat the process. Indispensable to his deliberations was an adequate engineering science of engine performance, hull drag, and other factors needed to make rational estimates. And for this, the traditional craft rules for building sailing ships were of little value, for nothing comparable had previously been attempted.

Brunel overcame multifaceted challenges in the design and construction of the *Great Western*. The vessel's hull was an unprecedented 236 feet in length, and two 750-horsepower steam engines, built to exacting standards, delivered power though elaborate mechanical linkages to the sturdily built pair of side paddlewheels.

His design succeeded. By steam power alone, the *Great Western* crossed the Atlantic from England to New York and arrived to a tumultuous welcome, and with a generous reserve of coal remaining unburned in the ship's bunkers.

With the *Great Western* hardly complete, Brunel launched a campaign to build an even larger steamship, the *Great Britain*, which he launched in 1843. Initially he envisioned a larger wooden hull, driven again by paddlewheels. Two technical innovations introduced on small coastal steamers, however, changed his mind. The first of these he observed on the *Archimedes*, a small steamship that was the first to use screw propellers instead of paddlewheels. With the idea of scaling up this form of power transmission, he had the *Archimedes* chartered to run tests on the propellers' efficacy. His experimentation culminated in a marine tug-of-war between two small coastal vessels, one propelled by a screw propeller and the other by paddlewheels, but in all other ways nearly identical. The propeller-driven craft was the easy winner, and Brunel abandoned side paddlewheels in favor of a screw propeller.

For the *Great Western*, Brunel had relied on traditional methods to construct a sturdy oaken hull. But wooden construction was problematical for the larger hull planned for the *Great Britain*. Doubling the dimensions of a hull increases not only its carrying capacity, but also its weight by eightfold. As Galileo had understood centuries before, however, the strength of the great beam that constitutes the keel would increase only fourfold. Thus for the *Great Britain*'s larger hull, even the strength of oaken construction would not do. When a single large ocean wave lifted the hull at its center, or when two waves supported the bow and stern of the vessel, while much of the ship's great weight hung suspended like a bridge across a crevasse, the keel would undergo more stress than oak could handle.

Brunel knew he was pushing the limits of wooden-hull construction. Meanwhile, his observations of the small iron hulls that were beginning to appear in coastal waters led him to consider wrought-iron construction. He concluded that iron's greater strength and lighter weight made it the material of choice for the 320-foot hull that he planned. And it would also eliminate the age-old problem of rot while ameliorating the increased fire hazards that came with replacing wind power with the burning of coal.

Effectively using iron, however, called for radical design changes. And construction of an iron hull of such size also brought unprecedented challenges, requiring the abandonment of much of the shipwrights' woodworking tradition. Wrought-iron plates were much stronger than wooden planking; hence the elaborate ribbing that was the foundation of wooded ship design no longer made sense, for the iron plating could now provide much of the hull's strength. Additional support was provided by adding bulkheads, walls of iron perpendicular to the ship's length that divide the hull into watertight compartments, thus compensating for iron's lack of buoyancy.

The switch from wood to iron caused turmoil in shipyard labor relations, for Brunel no longer relied heavily on the carpentry skills of the traditional shipwrights. The challenges were not in woodworking but in shaping and cutting iron plates and in riveting them together with watertight seams. The skills of boilermakers were much more relevant to his task, even though their previous employment included no experience in shipbuilding. The tradition-bound reluctance of shipwrights to learn the metalworking methods needed to fabricate the iron hull exacerbated the situation. Such dislocation is a theme often repeated, for whenever radically new technologies come into being, the new modes of production cause grave hardships for workers ill equipped to deal with change.

With each ship Brunel utilized design, construction, and testing methods that moved ever closer to science-based methodologies of today. Carrying over his earlier experience in railroad bridge building, he parted further from the craft tradition in making extensive use of the testing of components, materials, and concepts before and during construction. Components were mocked up and loaded, and he employed destructive tests to determine the limits of their material strength. Brunel's associate, astute naval architect John Scott Russell, worked on the theory of hull design and identified turbulence in the form of wave generation as a source of power dissipation. Brunel also consulted William Froude, who developed experimental methods, still used today, for towing scale models in a tank to determine the drag on the hull in terms of the towing speed and model shape and size.

The *Great Britain* succeeded in cutting the time to cross the Atlantic to fourteen days. Its crossings served as a prologue to Brunel's last and most ambitious steamship, the 692-foot *Great Eastern*. The *Great Eastern*, or *Leviathan*, as it was first called, was to have such carrying capacity that it could steam from England to Australia and return without refueling. It took iron-hull design to a new level of sophistication, including a watertight double hull and bulkheads that divided the ship into ten watertight compartments. The cellular structure of the double hull both provided protection should one or more of the watertight compartments be damaged and added to the hull's great strength. The ship's size required tremendous power to propel it at a reasonable speed, and for this, Brunel employed both side paddlewheels and screw propulsion. Launched in 1858, the *Great Eastern* was a tribute to Brunel's engineering acumen. Its hull was the largest to be built for many years to come, and its engines produced the largest concentration of power known at that time.

Between Galileo's beam analysis and the launch of the *Great Eastern*, more than 250 years elapsed in the struggle toward synthesizing the methods of science with the practice of engineering. Beginning with a slow start, the synthesis gathered momentum and came to fruition during the century corresponding to the industrial revolution. The results were dramatic, viewed from the outside in terms of the quickening pace of technological development, but also from within, in terms of the changed methodological approach of engineers to their profession. Fittingly, when engineers came together in Britain to establish their first professional society, the Society of Civil Engineers, they honored one of the great engineers of their time by referring to themselves as the Smeatonians and chose as their motto the apocryphal injunction from the Wisdom of Solomon 11:20: "all things in measure, number, and weight."

The complexity of the transformation from craft to science-based discipline is difficult to summarize. It may be helpful, however, to view it in terms of a redrawn version of figure 32 in which "Design" is inserted as shown in figure 38. To do justice to design, we must include a good many things. The expression of concepts on paper, which developed to a large degree in the Renaissance advances in drawing, of course, is at the heart of the process. First as sketches for examining conceptual alternatives, and then with increasing refinement and detail, the visualization process culminates in working drawings and specifications from which ideas can be implemented in the construction process.

But without engineering science—without the ability to analyze and to predict behavior before construction commences—the value of drawing cannot be realized to its fullest. Judgment gained from previous experience is invaluable, but the analysis wrought from the accumulated knowledge of the behavior of engineered artifacts must also be brought to bear. Thus it was the use of design rules, data banks of materials properties, and component testing combined with the successful application of scale models and mock-ups that transformed engineering. Advances in theoretical analysis and experimental method brought about the ability to make predictive calculations. Thus the Renaissance heritage of drawing and the legacy of the scientific revolution stand as cornerstones to modern engineering. Moreover, twentieth-century electronics allowed engineers to merge these cornerstones into something more: powerful computer-aided design and engineering methods have further facilitated the vital link between idea and artifact.

The development of scientific procedures also transformed testing from its crude origins in the trial-and-error methods of the preindustrial age. In modern practice a profusion of scientific instru-

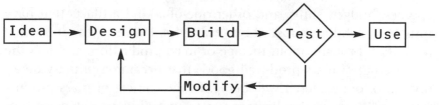

Figure 38. Design procedure. (Diagram by E. E. Lewis.)

mentation allows detailed validation of design calculations and confirmation of performance goals. Power, efficiency, strength, and durability are measured. Structures are loaded and engines stressed beyond their designed capacity in environments more extreme than those they are expected to encounter. Durability testing of components and systems indicates wear and life expectancy. Engineers use the results of such tests to modify and refine design, to confirm performance, and to assure adequate margins of safety and reliability.

The inseparability of design, construction, and testing from methods of scientific analysis and experiment is a hallmark of modern engineering. There are, however, two other distinguishing features of modern engineering for which figure 38 does not do justice. These are inadequately represented by "Idea" on the left and by "Use" on the right.

On the left of the diagram we have written "Idea" as the starting point for the engineering endeavor. That idea may be thought of as a conception of a material artifact to fill a human want or need. Ingenious inventors though the ages have provided such ideas. The source of creativity that gave birth to the spoked wheel, the mechanical clock, the sailing ship rudder, and other technological innovations of the preindustrial age will forever remain mysterious. However, from about the beginning of the nineteenth century scientific discovery began to provide the impetus for increasing numbers of new technologies. Scientific discovery as the source of new technology is here distinguished from the use of scientific methodology for the study, design, and testing of engineered artifacts. Scientific discovery of new phenomena and the impact that it has had on the engineering profession are topics we shall take up later.

Returning to Figure 38, we note that on the right we have simply written "Use." That may be adequate for the waterwheels, steam engines, bridges, ships, and other one-of-a-kind artifacts that have thus far constituted most of our examples. However, the industrial revolution brought about the age of mass production, and it is the engineering of mass-produced goods that became a primary determinant of our material well-being. In the context of mass production the "Build" in the diagram means to build a prototype or prototypes for use in testing, evaluation, and improvement. What happens next is then not immediate use but the implementation of technically and socially complex processes of manufacturing. The historic change from craft to mass production brought about both increased living standards and societal upheavals. In the next chapter we will examine mass production, its roots, and the immense impact it had on engineering and society.

Seven

Roots of Our Wherewithal

As I recently toured an automobile assembly plant, I thought, this truly is where it all comes together. As the engine and body of each car were successively lowered onto the chassis, the seats inserted, the wheels put on, and a myriad of parts and components installed, it rolled down the assembly line toward completion. The assembling car bodies passed the long sequence of stations at a rate in excess of one per minute. The entire automobile took shape from thousands of parts and components as it traveled from one end of the assembly line to the other.

Industrial engineers must deal with the massive technical and organizational challenges underlying the construction and operation of facilities as large and complex as an automotive assembly plant. They employ computer simulations—expert systems, neural networks, and other theoretical constructs—to select machinery and to optimize the sequencing of operations along the line. They must balance the line, specifying tasks performed by robots and those performed by workers to require comparable lengths of time. Otherwise, pileups would occur at some line locations while others sat idle. They study human-factors engineering to minimize worker fatigue and eliminate repetitive-motion injuries; they perform reliability analyses to reduce the number of machine breakdowns and the costly line shutdowns that result. Their planning must ensure that spark plugs, tires, windows, seats, batteries, and scores of other components manufactured by suppliers located across the country arrive just in time. Components must arrive not so early as to overwhelm the storage facilities but not so late as to cause the entire line to halt for lack of one small part. On top of all this, the engineers must design the production facility for flexibility so that each successive auto coming down the line can be assembled to order with the different combinations of colors, upholstery, and features that customers' tastes demand.

As in most manufacturing, highly automated machinery has replaced manual labor in performing many of the tasks in building automobiles. With a single wallop, gigantic presses at stamping plants shape flat sheets of steel into doors, hoods, trunks, and panels. Robots weld the panels together to form the car's body, such as shown in figure 39, and others then spray multiple coats of paint to

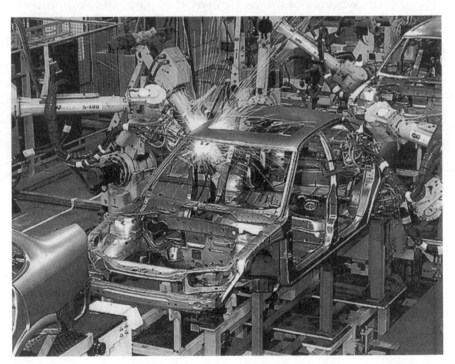

Figure 39. An automated assembly line.
(Reproduced by permission of the General Motors Archive.)

produce the sheen of the finish. Among the machinery there remain some tasks that only production workers could perform, tasks that require manual dexterity beyond what could be expected of a robot or that necessitate choices, decisions, or the exercise of judgment.

With each passing year, the hours of manual labor required to produce an automobile become fewer, and increasing numbers of the remaining production workers find themselves occupied with the more challenging pursuits of diagnosing problems, maintaining machinery, and preventing costly shutdowns of the line. The plant visit emphasized why industrial societies are evolving away from manufacturing to become service economies. For just as the earlier mechanization of agriculture drastically reduced the number of farmers needed to produce the nation's food supply, the highly mechanized assembly plant highlights the decreasing labor force needed to meet our needs for manufactured goods.

Between the craft shops of agrarian societies and these highly mechanized production facilities of our industrialized world lay the vast gulf we know as the industrial revolution. This revolution brought into being new technologies and economies of production that led in time to unprecedented increases in standards of living. It also brought immeasurable social changes, and for some, those changes were painful. The industrial revolution also transformed the nature of engineering, not only how engineers designed technological products but, even more, how they built them. An appreciation of these effects comes through examining more closely the paths of technical innovation and social change, the intertwined effects of specialization, mechanization, and the rise of precision manufacture that underlie the practice of engineering today.

A salient characteristic of modern manufacturing is the fine division of tasks. At the assembly plant I visited, workers performed a single task or small range of tasks that they repeated at minute-long intervals. Likewise, both totally automated machinery and that operated by line workers carried out very specific sets of functions, again repeated at one-minute intervals. This high degree of specialization of machines and manual labor is not unique to auto assembly lines; it is inherent to mass-production facilities manufacturing everything from toasters to trucks.

The division of labor, to some extent, dates back to the beginnings of civilization, to early Babylon, Egypt, and China, where agriculture first progressed to the point at which surplus production began to support urban populations. Soon after, potters, masons, weavers, carpenters, and other artisans began to cater to the more varied needs of the populace. From such early beginnings, highly refined crafts developed that later were to permeate Europe and remain in practice well into the industrial revolution. These crafts provided specialization by product. Like the wheelwright discussed earlier, each artisan became skilled in the varied sequences of tasks needed for the production and repair of the craft's products. Architecture, shipbuilding, and other larger undertakings required the cooperative effort of many craftsmen, frequently coming from a variety of guilds. Proud traditions developed around each of the craft guilds. Room remained for individualism, often expressed in unique ornamentation. And even though some specialization existed within the guilds, each worker performed a broad enough range of tasks that he could take pride in identifying with the finished product.

With time, such guilds often tended to become monopolistic, to maintain artificially high prices for their products, to resist change, and to impede technological progress. Against this backdrop, however, some master craftsmen sought the opportunity to improve the economies of production through divisions of labor. If a shop, for example, had orders to build not one wagon but several, and the master wainwright employed several craftsmen, then it made sense to separate the construction tasks beyond the traditional bifurcation between the wheels and wagon chassis. He could have workers specialize on various parts of the chassis, on the axles, or on the wheel spokes, rims, and hubs. Or perhaps he could organize according to the type of woodworking, separating cutting and drilling from shaping and finishing. A worker specializing in each of these fabrication steps was bound to become more adept at performing a narrower range of duties. Likewise, others could hone their skills solely on fitting the pieces of the wheel, axle, or chassis together and working with the smith to secure the wheel rims and other metal fittings. With sufficiently refined specialization, the master could employ workers with less extensive skills. They no longer required a lengthy apprenticeship to become proficient in all facets of the trade. Instead, the master, or other workers, could more quickly provide them with enough training to gain the manual dexterity and judgment required to perform a narrow set of operations.

In such a shop, the master's activities would become increasingly divorced from those of his employees as the tasks became more finely divided. He devoted more energy to managing the enterprise and marketing its products than to plying his craft. The worker, unlike the master, had no need to be involved in soliciting business, deciding what should be built, securing raw materials, or selling the finished product. The master's decisions would determine what was to be built, and most of his contact with the worker would be spent in instruction, supervision, and coordination of the labor of others. If the enterprise grew, the master profited handsomely. The more successful master craftsmen often became well-to-do capitalist merchants, whereas the less successful fell into the ranks of the specialized workers, laboring for wages in the shops of others. Thus it was that master craftsmen, merchants, and others gained sway over enterprises, as divisions of labor led to more economical and more profitable production of goods. The traditions of craft production succumbed to early forms of capitalism and to the precursors of modern methods of mass production.

Early examples of successes in dividing tasks to increase production and lower its costs appeared long before the industrial revolution. Ancient Greeks understood the economics of dividing manufacture into narrow tasks. Greek historian Xenophon observed, "One man earns a living by only stitching shoes, another by cutting them out, another by sewing the uppers together, while there is another who performs none of these operations but only assembles the parts."[1]

In pre-Renaissance Italy, the division of labor in textile manufacture became widespread. The merchants of Florence divided cloth production into more than twenty-five distinct steps, each performed by workers who specialized only in that operation. The costs of producing the finished product fell. And the resultant profit stemming from these productivity gains contributed considerably to the concentration of wealth that led to the city's Renaissance flowering.

Early in the industrial revolution—before the momentous changes brought about by mechanization, the steam engine, and iron construction—capitalist entrepreneurs became increasingly aware of the efficiencies to be gained by aggressively pursuing finer divisions of labor and reasoned layout of manufacturing facilities. The production of ceramic dishes and cookware, carried out by potters using methods handed down from antiquity, was an important industry that offered such an opportunity. And Josiah Wedgwood seized it. He set out to reduce the cost of making pottery while improving its quality. He accomplished this, without the need for new technologies, by radically reorganizing the means of production. He separated the making of pottery into finely divided tasks, and subjected his workers to levels of regimentation unheard of in craft production.

Wedgwood pioneered the factory production of pottery at Etruria in England. He first divided his plant by the class of ceramics to be produced: useful, decorative, and so on. Within each product class, he carefully organized his operations through fine subdivision of the tasks required for production. He began with the selection, grinding, and mixing of the necessary materials. He continued through forming and turning of the articles, the kiln baking, and the final decorating and finishing. He organized the flow of work and finely divided the tasks so that each laborer had a narrow specialty: grinder, mixer, dipper, brusher, molder, turner, oven-fireman, coal-getter, painter, enameller, and gilder. Some tasks required more skill than others, and that led to further economies from the division of labor, for skilled workers did not waste a part of their time performing the more menial tasks. But more importantly, each division of labor reduced skill requirements, made training easier, and enlarged the pool of available workers.

Wedgwood laid out his plant such that workers first formed the pots and then passed them to the paint room, then to the kiln room, and successively to the account room, where an inventory of the production took place. Finally the pots were passed into storage. Unlike earlier potteries, where workers wandered from place to place as well as from one task to the next, Wedgwood assigned each worker to a particular post to accomplish a single task. By 1790 only five of Wedgwood's nearly three hundred workers had no assigned post; all the others were specialists assigned to a single task. Not only did specialization drastically cut the costs of production, it also resulted in improved quality of the finished product. Wedgwood's pottery was superior to that produced by his competitors.

Like Wedgwood, others sought to rationalize plant layout and work procedures—to reduce duplication, wasted steps, and other inefficiencies. Eighteenth-century economist Adam Smith described the culmination of this division of labor. In *The Wealth of Nations*, he set forth the economies to be gained by specialization in his description of producing something as simple as a pin: "One man draws out the wire; another straights it; a third cuts it; a fourth points it; a fifth grinds it at the top for receiving the head; to make the head requires two or three distinct operations; to put it on is a peculiar business; to whiten the pin in another; it is even a trade by itself to put them into the paper; and the important business of making a pin in this manner divided into about 18 distinct operations."[2] Indeed, as the industrial revolution progressed, ever finer divisions of labor took hold. Minimal training replaced the need for extended apprenticeships, allowing entrepreneurs to hire unskilled workers to perform the finely divided tasks.

The acclimation of workers to factory employment was a severe challenge, for it subjected them to regimentation and discipline that was alien to their previous experience. Some of those recruited for factory employment were workers displaced from vanishing craft occupations. Far more frequently, however, impoverished agricultural laborers—men, women, and children no longer able to eke out a living on the land—sought to improve their material well-being through wage labor. No one can doubt the long and arduous hours they had previously toiled in the field or the shop. Drudgery may have dominated their existence, but at least they had some control over the pace at which they worked. Their workday had not been determined by a clock but by the varied tasks that they needed to complete. Workdays had been long but were varied by task and with the hours of daylight as the seasons changed. Some variety also existed in the tasks they performed, if not from day to day then at least from season to season. Agricultural laborers often practiced simple but useful handicrafts when the weather kept them out of the fields, and artisans frequently cut back the hours devoted to craft production to participate in planting or harvesting.

In contrast, factory discipline of the industrial revolution subjected workers to twelve-hour workdays with one short meal break. Wage laborers performed highly repetitious, mind-numbing tasks at an inalterable prescribed pace, a pace necessary to maintain coordination in the workflow, and—as mechanization increased—to optimize machine usage. Training workers was simple, and dividing the tasks and laying out the workflow were manageable undertakings. Far more challenging was maintaining the novel form of discipline imposed on workers to achieve the coordination required to accomplish the highly organized tasks. Wedgwood struggled with these problems, and arguably his most significant triumph was in instituting a hierarchy of supervisors and managers to enforce sobriety and punctual attendance and to reduce absenteeism.

Labor issues brought forth by the factory system continued to challenge efforts to organize and manage manufacturing processes to ever-higher levels of efficiency. More than a century after Josiah Wedgwood's accomplishments, American efficiency engineer Frederick Winslow Taylor became a widely recognized and highly influential advocate for a systematic approach for improving worker productivity. In his text *The Principles of Scientific Management*, he set forward methods that constitute the early attempts to bring forth a rational practice of production engineering.[3]

Taylor established the field of time and motion study. He examined common manual processes performed by laborers, the loading of pig iron being the most frequently cited example. Through careful observation he determined how a given operation could be performed with the least amount of effort, eliminating wasted or unnecessarily strenuous motions. He instructed workers on what he considered to be the correct way to perform their tasks; he also adjusted table heights, walkways, and other physical features of the workplace in order to make the task less strenuous and therefore more efficient.

Taylor also carried a stopwatch to time the workers' motions, for his criterion for increased efficiency was that the task took less time to complete. His time studies, examining how fast a worker could accomplish a task, were part of a larger management effort to measure labor and to determine what constituted a reasonable day's work. In Taylor's view, such studies were beneficial to the workers, since he believed that workers' sole motivation was economic gain. Thus if he showed them how to increase the speed at which they performed their assigned tasks without increased effort and paid them on a piece rate basis, the result should be beneficial to all.

But worker suspicion and resistance to Taylor's methods grew. His narrow economic view of their motivation did not consider powerful psychological needs such as camaraderie with fellow workers and a sense of identity with the products to which their toil contributed. Only later, as the understanding of industrial psychology advanced, did industrial engineers address such human considerations. Moreover, workers rightly questioned how the rewards of their increased productivity were to be divided between management and labor. With stopwatches and tape measures, efficiency experts expounding Taylorist principles often found themselves embroiled in worker-management struggles. The "speed-ups" stemming from their studies often seemed advantageous only to management, leaving the workers only with a more stressful workplace.

Another powerful development also eclipsed Taylor's work. His studies focused on the speed and effectiveness of manual labor. By the late nineteenth century, when Taylor was performing his experiments, setting the pace of production—based on a "scientifically determined" rate at which manual labor could be performed—was becoming a moot question. Mechanization of production processes had begun more than a century earlier in England's textile industry, where inventors succeeded in using combinations of bobbins, rollers, and other mechanical contrivances to mimic the finger and arm motions of spinners as they operated their spinning wheels. Beginning in the 1760s, the succession of machines first for spinning yarn and later for weaving and for cleaning wool transformed textile production. What had been a labor-intensive handicraft industry became a machinery-driven factory system, producing textiles that were both cheaper and of higher quality.

Mechanization propelled the industrial revolution forward, with newly invented machinery spreading from textiles to other industries and bringing with it dramatic productivity increases. But mechanization also spread the need for factory discipline. This discipline differed significantly from that of Wedgwood's pottery plant or the manual tasks studied by Taylor. The sequence of operations and the rate at which manual labor could accomplish them were less important than machinery characteristics in determining factory organization. The machinery replacing hand methods was expensive to build, and it operated continuously, consuming energy resources. Its effectiveness necessitated a more rigid time discipline in factory work. Machine-paced labor became the rule for those who kept the machines supplied with material, removed the finished products, and made adjustments to the running machines. Industrial engineers increasingly had to deal with the challenges of the human-machine interface. These challenges were and continue to be pivotal in realizing the immense potential of mechanization for increasing living standards and for securing tolerable working environments in the factories.

Giant punch presses and stamping machines; precise computer-controlled lathes, grinders, and milling machines; and highly sophisticated robotic manipulators—all these and more are central to the economic production of automobiles, appliances, and the host of other mass-produced products of our daily lives. The high degree of mechanization is as essential as the divisions of labor and factory organization to the industrial productivity that we take for granted.

We may rightfully associate the beginnings of mechanization's explosive growth with the industrial revolution of the late eighteenth and nineteenth century, but it has roots that go back much farther than that.

Since long before the industrial revolution, millwrights had been building power machinery—waterwheels and windmills—of increasing sophistication. Their machine-building skills expanded as they learned to build blast furnace bellows, pumps, trip hammers, and other machinery for converting the energy of wind, water, and animal muscle to practical use. Wood was the construction material for this labor-saving machinery, as it also was for the ships, barges, wagons, and coaches that served as the primary means of transport. Until steam power and cheaper iron became available, the limited strength, uniformity, and durability of wood severely curtailed what could be engineered. Equally important, the limiting properties of wood precluded the precision manufacture needed for cutting gears, screws, and other intricate shapes essential for creating the more powerful and useful machinery that came with the industrial revolution.

Modern mechanical machinery also has roots in a second craft tradition. Since the fourteenth century, clocks have been an integral part of European culture. The clock mechanism was an ingenious invention that spread from medieval monastic buildings to the town halls across Europe, then to the residences of the wealthy. By the sixteenth century, clock makers were constructing not only elaborate clocks but also portable watches with spring-driven mechanisms. Clock and watch making required extraordinary craftsmanship in working brass and steel to create these intricate and ingenious mechanisms. They formed, cut, finished, and polished the intricate interlocking parts accurately and often engraved and decorated them with inlayed jewels as well.

These metalworking skills expanded to encompass musket gunlocks and other intricate mechanisms. Increasingly, mathematics was combined with the clock makers' skills to build devices for surveying and navigation, from which the instrument makers' trade emerged. Consequently, those who designed and built such instruments had to acquire knowledge of mathematics far beyond that required of other craftworkers. The rise of scientific investigation in the seventeenth and eighteenth centuries further increased skills in mechanical design and construction, as wood and other construction materials were replaced by brass and steel components as the need for precision grew. By the eve of the industrial revolution the instrument makers had mastered the technologies of precision machining: the cutting of gears and screws and the fabrication of intricate mechanisms of gears, levers, ratchets, springs, and other mechanical elements. They learned to select metals for increased durability. They designed mechanisms to use lubricants effectively, to reduce the effects of friction, and to compensate for the effects of temperature change on materials' properties. Instrument making was a springboard from which the mechanical engineering of the industrial revolution emerged. It should be no surprise that James Watt, John Smeaton, and other notable engineers of the eighteenth century began their careers as instrument makers.

Concomitant with the transformation from craft to machine production, the construction of the machinery itself evolved rapidly. The frames and larger components of early machinery, such as that which mechanized the spinning of yarn, were wooden, whereas the iron parts, made by blacksmiths, were few. Increasingly, clock makers and instrument makers entered the machine-building trades to construct the more intricate gearing, springs, and linkages from brass and steel. Demand grew for strength, rigidity, and durability in textile machinery and even more so in steam-driven pumps as well as drilling, hoisting, and other steam-driven equipment in what came to be known as the engineering industries; however, these were qualities that wood structures and brass gearing could not provide. The demand was to be met eventually by constructing the machinery from iron.

Several developments played major roles in bringing about the widespread use of iron. Perhaps most important, in the early eighteenth century, Abraham Darby invented the coking process. Whereas previously iron had been smelted from its ore in small amounts using charcoal obtained from wood, Darby's process obtained coke from coal, and coke was an ideal charcoal substitute.

Thus by substituting plentiful coal for increasingly scarce wood as the raw material used in smelting, coking brought down the cost of iron, and allowed it to be produced in much larger quantities. The supply and the demand for iron increased hand in hand. The development of the steam engine both demanded iron for construction and produced the greater quantities of power needed for the blast furnaces and rolling mills to smelt and form larger quantities of iron. As demand for iron grew steadily in the nineteenth century, advances in metallurgical methods, first the puddling process and then the Bessemer converter, further increased the economical supply of wrought and cast iron and, later in the nineteenth century, steel. The larger quantities of iron found immediate use in the rails of expanding railroads, and for the beams, columns, and cables of bridges, buildings, and other structures. Construction of iron machinery, however, was more problematic.

Shaping and cutting iron is much more difficult than working with wood, and hand tools—adequate for the woodworking crafts— were of little value for producing larger gears, shafts, and other machine elements from iron. The ancient trade of blacksmithing, geared as it was to forging small iron implements, was totally inadequate. Likewise, the precise brass-working techniques of instrument makers proved insufficient for the harder metal, and those of the gunsmith and other artisans were limited to very small iron components that could be shaped with the chisel, hammer, and file. The construction of machinery required much larger iron components; forming and cutting them required tools exerting much larger forces than artisans' muscle could provide.

Pioneering builders of the nineteenth century met this challenge by inventing numerous machine tools. Powered by a machine— most frequently a steam engine—these tools exerted the force and precision needed for forging, drilling, cutting, grinding, turning, and polishing iron gears, shafts, and bearings needed for machine construction. With the development of metalworking lathes, grinders, drills, and other machine tools, the blacksmith's and instrument maker's trades were subsumed into the machine shops that became the backbone of nineteenth-century mechanical engineering projects. These shops produced everything from the machinery for spin-ning and weaving cotton and woolen fabrics to the engines that powered railroads and steamships.

Machine tools required a shop organization that differed greatly from the craft shops of old, where artisans had moved from place to place, leaving the their hand tools idle. Steam engines required large capital investments for construction and large expenditures for the coal that kept them running. Moreover, efficiency demanded contin-uous operation, wasting as little fuel as possible in the cooldown and heat up that accompanies each shutdown and start-up. Thus work discipline required that tenders keep their machines continu-ally active, for nearly as much energy was dissipated when the machine tool was idling as when it was in active use.

The friction of mechanical linkages—the shafts, pulleys, gears, and belts—dissipated energy as they distributed power from the steam engine to the lathes, drills, and other metalworking machinery. To minimize frictional losses, the steam engine was most commonly placed in the basement of a multistory factory building. Power was transmitted off the engine's large flywheel vertically up through the building. It traveled through shafts and belts to secondary shafts, which in turn transmitted it by pulleys to individual machines. The factory's designers placed the machinery with the largest energy con-sumption closest to the basement engine room and crowded the remaining machinery as close together as possible to minimize trans-mission losses. Consequently, the workers found themselves squeezed into very cramped quarters, such as those illustrated in figure 40.

Steam-powered mechanization presented a dilemma for further increasing manufacturing productivity. Logical plant layout for the effective flow of work and materials improved productivity. Indeed, such logic was at the heart of Josiah Wedgwood's improved pottery production. But logical plant organization succumbed to energy con-servation requirements in steam-powered production of metal prod-ucts. The solution did not come until late in the nineteenth century: it was the electric motor, based on Michael Faraday's discovery decades earlier that changes in a magnetic field induced a current into an electrical circuit. The motor provided a versatile new method

Figure 40. Mechanical energy transmission at the Weed bicycle factory, 1881.
(Image from Smithsonian Institution.)

for powering not only the machine tools of the metalworking industries but for virtually all other types of machinery as well.

Shortly after the turn of the twentieth century, the use of individual electric motors to run separate pieces of machinery became widespread. The motors freed the placement of machinery from the constraints of the shafts, pulleys, and belts that had transmitted mechanical energy from a centrally located steam engine to all the machinery in a shop. The cramped quarters needed to minimize transmission losses were no longer necessary, for the electrical transmission of energy to the machinery—even over long distances—entailed practically no power losses. The electric motor thus freed industrial engineers to lay out the factory to facilitate the flow of materials and to organize the placement of machinery such that operations could be sequenced in logical progressions. The stage was thus set for combining mechanization with factory organization in a more logical manner. This was to come to fruition in the assembly lines of the twentieth century.

Before the synergy of mechanization and factory organization could bear fruit, however, engineers faced another challenge. Whether machine shops fabricated parts for locks, clocks, pumps, guns, or engines, those parts had to be fit together to form working mechanisms. The procedure was long and tedious, requiring high-skilled metalworkers to grind, file, and chisel the machine-made parts until they mated properly to form a working whole. The challenge was to eliminate such laborious fitting by producing parts with such precision that they were interchangeable and could simply be assembled, without customized shaping.

❋   ❋   ❋

The modern production plant that I toured could not exist without exceedingly precise manufacturing of each of the component parts installed in the autos as they progressed down the assembly line. At each of the line's stations the appropriate parts, mechanisms, components, or subassemblies were taped, snapped, bolted, or riveted into place. Each came from an inventory of identical parts, so alike that they fit interchangeably into place. The availability of parts fabricated with precise dimensional accuracy is a cornerstone of present-day mass production. Components fabricated at remote sites must be assembled without the need for custom fitting. Tolerances must be tight enough that the parts are interchangeable. Likewise, present-day maintenance and repair services presume the availability of interchangeable repair parts. If a part breaks or wears out, a supplier provides the repair shop with an identical repair part; the shop doesn't custom build it.

Interchangeable parts manufacture did not exist until late in the industrial revolution. How it came into being is closely intertwined with systematizing and mechanizing production, but it followed a quite different path from the transformation from craft to mass production of textiles, cookware, and other essentials of everyday life. Entrepreneurs in search of profits saw lucrative markets if they could reduce the costs of producing those widely used products. But they saw no comparable profits waiting in producing clocks, instruments, gunlocks, and other precision metal mechanisms for which the demand was much more limited. Moreover, gunsmiths, clock makers, and instrument mechanics who custom crafted such valuable items were among the most highly skilled and respected artisans of the time. Entrepreneurial attempts to reduce these crafts to the factory discipline needed to achieve the interchangeable parts precision would prove difficult. It presented challenges not present in displacing spinners', weavers', or potters' crafted products with inexpensive, factory-made substitutes. As a result, eighteenth-century capitalists left the quest for interchangeable parts manufacture to others. Farsighted military engineers championed precision manufacture, for they foresaw tremendous advantage to mass-producing muskets and other weaponry so nearly identical that their parts were interchangeable.

We may trace early attempts to manufacture parts interchangeably to eighteenth-century France, where engineering theory was most advanced and where formal engineering education first developed. Rigorous preparation gave rise to an elite cadre of military engineers who were determined to systematize not only the execution of war but also the design and the manufacture of weaponry. These Enlightenment engineers insisted on more accurate and reliable muskets than those crafted by the gunsmiths of the day. They set out to produce tens of thousands of muskets that were identical, or so nearly so that they could be assembled from interchangeable parts.

The engineers' rationale was clear. Precision manufacture with interchangeable parts offered great battlefield advantage. Uniformity would improve weapon accuracy, and when a musket was lost or damaged, without delay the musketeer could effectively fire another. Battlefield repair would be simplified immensely, for military men could be trained to perform parts replacement on the spot, obviating the need to ship damaged weapons back to the armories to be rebuilt. And in the longer term, making the manufacturing process more uniform would lead to sharp reductions in production costs.

Imposing uniformity on the manufacturing process was a daunting objective, given that production took place at several armories with locations widely dispersed across France. Moreover, success required drastic changes in the craft ethos of the gunsmith. Traditionally, a smith crafted the entire weapon—lock, stock, and barrel, as the saying goes. Of these the gunlock, or firing mechanism, consisting of a number of interlocking small metal parts, required the greatest skill. The smith first forged the individual parts. But forging was not nearly precise enough for the parts to fit together to form a working gunlock. Rather, the gunsmith spent hours chiseling, grinding, and filing the parts to their proper dimensions and shape until at last he could fit them together to form a working mechanism.

No two weapons crafted in this manner were exactly alike; moreover, firing characteristics varied substantially from musket to musket. Thus the musketeer expended considerable time and effort learning to account for his weapon's firing idiosyncrasies before he could use it effectively. If one of the gunlock's parts broke or wore out, the weapon had to go back to the gunsmith. The smith in turn forged a new part and ground and filed it and its mating parts until he once again succeeded in fitting the mechanisms back together and putting the lock in working order. Invariably, the repair process altered the weapon's firing characteristics, requiring the musketeer's considerable effort to adjust to the reconstituted weapon.

Improving musket uniformity was no simple task, for the production methods envisioned by the military engineers were in sharp contrast to the experience of the artisan gunsmiths. The gunsmiths had worked hard but not in a systematic regimen. They intertwined their work with social intercourse. The working day depended on their dexterity and upon the pace at which they chose to work. The tasks to be accomplished varied from day to day, and the length of the day varied with the tasks to be completed.

The engineers sought to establish a disciplined workforce with substantial divisions of labor. In the extreme, their approach assigned a single part to each smith to produce over and over. An inspector then measured critical dimensions of each part. If it was not within tolerance, he rejected the part, and the smith was not paid for the piece. Only with such discipline did the engineers believe that interchangeable part production could succeed and the time-consuming process for fitting the parts together eliminated. The artisans' guilds chafed under such regimentation and work discipline, and many struggles ensued as the engineers attempted to subjugate the craft workers' autonomy to objective measures of performance. No doubt, the dehumanization of their labors also took its toll on the workers' psyches.

Some success at greater division of labor ensued under the auspices of the French military engineers. Honoré Blanc divided the production of gunlocks into over 150 distinct operations and invented special-purpose tools to mechanize a number of the tasks. He devised fixtures and jigs to hold the work pieces in place and gauges to measure the accuracy of the outcomes. The resistance of the artisan guilds notwithstanding, he succeeded in producing interchangeable parts for the muskets' flintlock mechanism. On the eve of the French Revolution, Blanc's success was brought to the attention of the US ambassador, Thomas Jefferson. In 1785 Jefferson visited Blanc's manufacturing facility located in a former dungeon at the Château de Vincennes. Blanc sorted the springs, tumblers, lock plates, and other parts for fifty gunlocks into bins. To Jefferson's amazement, he quickly assembled a number of working gunlocks from parts drawn from the bins. The demonstration so impressed Jefferson that he had several of the gunlocks shipped to the United States, and following his return he advocated the virtues of interchangeability.

Following the French Revolution, interchangeable parts manufacture languished, falling victim to the sustained opposition of the craft guilds, the entrenched practices of arms merchants, and the changed political climate. The idea took root, however, in the United States. Interchangeable manufacture fired the imagination of Eli Whitney. Famous for his invention of the cotton gin and its revolutionary impact on southern agriculture, Whitney became a highly visible champion for uniform parts production, proposing to Congress a revolutionary system of mechanized manufacture for producing military muskets. Financed by government contracts, he established a private arsenal at Rock Mill, Connecticut, where he attempted to implement a system for precision musket manufacture. Although his effort drew a great deal of attention, his successes were limited, and it is doubtful that he ever approached the production of truly interchangeable parts. It was, rather, the sustained efforts of the US Army Ordnance Department that eventually brought about the technological innovations and manufacturing organization required to establish true interchangeability in weapons production.

In the early decades of the nineteenth century, officers educated at the newly established US Military Academy at West Point staffed key positions in the Ordnance Department. The curriculum at West Point, the first engineering school in the United States, was closely patterned after the École Polytechnique and earlier elite engineering schools of France. Thus US officers were well versed in engineering theory, and like their French counterparts, they strove to systematize the design and production of weaponry. They, too, appreciated the potential of interchangeability for weapons uniformity and accuracy, for speedy repair on the battlefield, and for cost savings.

At the federal armories at Springfield, Massachusetts, and Harpers Ferry, Virginia, the Ordnance Department sponsored the innovations of able mechanical inventors and pressed for the changes in arsenal organization necessary for interchangeable parts manufacture. As in France, gunsmiths resisted giving up the autonomy that the age-old craft methods afforded them and frequently sought to undermine the imposition of the work discipline and divisions of labor. But in the Untied States the situation was more conducive to change than in Europe. Labor was in short supply in the young and expanding country, and the guilds' apprenticeship requirements and restrictive admissions standards were not so powerfully entrenched. Rather than resist the regimentation of their work environment, craftsmen frequently abandoned their guilds to move west in search of greater opportunities. The aggravated labor shortages lessened opposition to employing less skilled workers to perform the simpler repetitive tasks, and the less skilled were content to have steady employment.

US engineers sponsored development of machine tools of increasing precision. They increased the division of labor and constructed mills, drills, and lathes devoted exclusively to one operation, each operable by laborers with minimal training. Over a ten-year span they divided weapons production at the Springfield armory into more than one hundred distinct operations. They progressively introduced precision jigs, fixtures, and gauges, such as those shown in figure 41, to accurately position the metalwork pieces in the machines and to measure their dimensions. Through such measures the armories gradually succeeded in producing parts of such dimensional accuracy that they were interchangeable: they did not require customized hand-filing and chiseling to fit them together.

At first, the drive for uniformity did not reduce production costs; in fact, it caused them to go up. Since private entrepreneurs foresaw little profit in such an enterprise, only the military persevered. To the Ordnance Department the higher cost of weaponry was more than offset by the increased performance, reliability, and ease of battlefield repair that interchangeability afforded. But in the longer term, refinement of the new methods tended to bring costs down. The skilled fitters who filed, bent, chiseled, and hammered the pieces together were replaced by less skilled assemblers who simply snapped or bolted them together, and in substantially less time. Thus the labor cost per weapon gradually fell well below that of craft production. Growth in the production rate then gave rise to increasing economies of scale. For if the costs incurred in designing and constructing the specialized machinery and in organizing the operations were prorated over large enough numbers of muskets, the total cost per weapon would also fall below that of craft production.

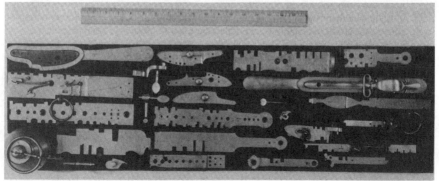

Figure 41. Rifle inspection gauges from the
Springfield federal armory, manufactured in 1841.
(Image from National Museum of American History, Smithsonian Institution.)

By the 1840s the benefits of interchangeable parts manufacture had been demonstrated at the federal arsenals, and as costs came down the techniques quickly spread to the private arms industry of Connecticut. The London Crystal Palace Exhibition in 1851 brought international prominence to the US firearms industry. The quality and precision of the revolvers, rifles, and other machine-made products drew much acclaim, and the methods developed in the Ordnance Department's arsenals became known as the "American System" of manufacture. So impressed was the British military that it sent an investigatory commission to visit the New England armories and in 1853 made wholesale transport of machine tools back to Britain to equip its new armory at Enfield, near London.

In the 1850s other industries began to realize the benefits of the American System to manufacture more uniform parts for sewing machines, watches, clocks, and locks. Implementation in the United States surged ahead of Europe, where age-old traditions of the metalworking crafts were not so easily displaced. The following decades brought these innovations to typewriters, agricultural implements, bicycles, and a host of other products. Even as machine-made products became increasingly uniform, however, companies frequently encountered difficulties; long periods elapsed before the filing and grinding of fitters succumbed completely to the goal of truly interchangeable assembly. Nevertheless, the American System, or armory practice as it was also called, resulted in rapid technical progress in the manufacture of complex metal mechanisms and became a cornerstone of the mass-production industries of the twentieth century.

The auto assembly plant I visited represents the culmination of manufacturing progress in one of the world's most important industries, an industry where manufacturing was transformed from craft to mass production and beyond during the course of the twentieth century. Before the turn of the century making automobiles—or horseless carriages, as they were then called—was a craft industry. In the United States and even more so in Europe, many small companies built automobiles in small numbers. Typically, a customer or client discussed the car's design, styling, performance, and likely cost with the designer. When they reached agreement, the designer drew up plans, leaving many of the details to the discretion of the skilled machinists assigned to build the car. They worked in moderately mechanized shops. Although primitive by today's standards, the lathes, grinders, drill presses, and milling machines of the day had already supplanted much of the work previously carried out with hand tools. The machinists fabricated parts with only moderate degrees of precision, since later they could file and chisel the individual parts to make them fit together to form a satisfactory whole. Once completed, the auto was taken on a test run, and the builders made further adjustments, and filed and chiseled some more, until finally it performed well enough for delivery to the buyer.

The high levels of craftsmanship characteristic of this process most often resulted in autos capable of good speed and comfortable ride. Styling was attractive, frequently to such a level that the autos became collector's items, preserved to this day in museums and private collections. But no two were alike. Even if the artisans made them from the same design, the amount of discretion left in the construction details coupled with the lack of interchangeable parts manufacture meant that each auto required individual tuning. Moreover, as time passed the autos needed frequent tuning and adjustment to maintain the performance achieved on the test run. And if a part should break, no mass-produced repair parts existed; a new part was fabricated to replace the old.

The auto's most salient feature was its high cost of craft production. Automobiles were luxury items, affordable only for the wealthy. And they were expensive to operate as well as to buy. Typically, the owner hired a chauffeur to drive the car, attend to the frequently needed adjustments, and see to it that parts were fabricated in a timely manner to replace those that had broken or worn out. Autos remained luxury possessions of the privileged few, until one of the more inventive builders envisioned combining design economies with methods for reducing the production costs to the point where the automobile would become affordable transportation for the average family. In 1907 Henry Ford set upon his quest to create the Model T.

Building on nineteenth-century manufacturing advances, Ford transformed automobile manufacture from a craft to the production methods that more closely resemble those of today. The power, precision, and durability of milling machines, lathes, and other machine tools available to him dwarfed those of a hundred years before. By the turn of the twentieth century, industrialists had established the American System and demonstrated the value of interchangeable parts production in manufacturing firearms, sewing machines, bicycles, and agricultural implements. By that time, industrial engineers were also free to lay out factories to facilitate the flow of work and materials, without regard to the proximity of a centrally located steam engine. For as discussed earlier, electrical generators were replacing steam engines, and energy could be transmitted nearly loss-free through electrical cables, allowing machinery powered by electric motors to be located wherever the engineers saw fit.

Ford established a culture of innovation in the manufacturing process. With his encouragement, engineers performed experiment after experiment to make production methods more efficient. And when they found a better way, the cost of scuttling old processes and tools didn't deter them. They split the steps needed to fabricate automotive components into a sequence of ever-more-finely divided tasks, allowing workers to become proficient quickly at their narrowly defined assignments. They also redesigned the machine tools accordingly. A general-purpose machine tool, such as a milling machine or lathe, required a skilled machinist to set it up, adjust it, and operate it over the range of tasks it could perform. The engineers replaced these with a profusion of single-purpose machines. They designed and preset each of these to perform one particular operation. They equipped each machine with fixtures to correctly position the work piece and securely hold it in place while machining took place. This all but eliminated the need for skill or judgment on the operator's part, dramatically reduced the time required for inserting and removing the work pieces, and resulted in more parts produced with fewer hours of labor.

Concomitantly, Ford's engineers vigorously pursued manufacturing precision. They insisted on high accuracy in all critical part dimensions to assure interchangeability and facilitate ease of assembly. Following Ford's lead, they pushed for levels of precision that allowed novice assemblers to replace the highly skilled fitters still required in many industries of the time. Other benefits to industry accrued as well. More uniformly manufactured autos performed more uniformly. Precision production curtailed and eventually eliminated the laborious test-and-fix process, in which workers drove and adjusted each finished auto until it performed adequately. And finally, precision production brought about the availability of interchangeable repair parts. These, Ford surmised, allowed owners to make many of their own repairs, sparing them the expense of securing the services of highly skilled specialists.

Ford approached Model T design and production as an integrated whole, instituting what today we would call "design for manufacture." His engineers paid attention to how design refinements affected the costs of production. Accordingly, they designed for low production cost. Ford kept the engineers working away at this, simplifying the design, combining two parts into one wherever possible to reduce fabrication costs and eliminating steps in the assembly process. They also redesigned parts to reduce the number of machining operations and to simplify assembly. On a larger scale, they worked to optimize the placement of the engine, drive train, and other Model T components both to speed the assembly process and to facilitate repair.

The company's engineers pioneered new methods for refined auto design, for parts fabrication, and—perhaps most revolutionary—for the organization of the assembly process. Instead of assigning one or a few men to assemble an entire automobile, as had been the craft custom, they had assemblers go from car to car in sequence, bringing their tools and parts with them, each attaching one or a few parts. Then they envisioned the even greater efficiencies to be gained by moving the autos to the workers and their tools. Thus Ford's greatest innovation, the assembly line, was born.

Ford's ideas drew on observation of the meat-packing industry, where overhead trolleys conveyed carcasses from worker to worker at a steady pace, as each made one particular cut in the butchering process. This was a true assembly—or rather disassembly—line, with stationary workers concentrating on a single task. Translating the concept from butchering hogs with hand tools to mechanized auto production with its many complex interlocking parts of metal, glass, and rubber required numerous innovations. But Ford's engineers took on the challenge. They first designed an assembly line for one component of the auto, the magneto flywheel, and it began operation in 1913. A remarkable reduction in flywheel assembly time, from twenty to five minutes, was the result.

The Ford engineers next expanded the line organization to take on a much larger task, the assembly of the chassis—the automobile's frame, wheels, and engine. Their efforts reduced the twelve-and-one-half-hour assembly time, required when parts were carried to a stationary assembly point, to six hours. The assembly line was crude, using a rope to pull the chassis past worker stations that were stockpiled with components, but improvements quickly followed. A chain drive to power assembly line movement replaced the rope, and the stations at which workers were located were redesigned for convenience and comfort. And by April 1914, chassis assembly time fell to one and a half hours, further reducing costs.

The complete assembly line, fully installed by 1915, accompanied by other manufacturing advances and continuing refinements of the Model T's design, caused production to skyrocket. In the early 1920s more than a million identical cars rolled off the assembly line each year. Peak production surpassed two million autos per year, and from the time that the Model T was first introduced, Ford produced over fifteen million of them. During the same time period, large and growing economies of scale in mass production, compounded by the ever-increasing production efficiency, allowed Ford to reduce prices steadily. This led to a two-thirds drop in the cost to the consumer from that of the first Model Ts. Ford's drive to mass-produce a car that the average family could afford met with great success. Profitability of the company flourished. Additionally, Ford acted on his conviction that mass production would be successful in the long run only if workers could afford the products they built. He offered his employees outstanding wages for the day.

Ford paid his assembly line workers five dollars per eight-hour working day. At the time, these wages were nearly double the going industrial rate for manual labor, and for a shorter workday. Nor could the landless hope ever to approach matching these wages for agricultural labor. Ford recruited former agricultural workers, whom farm mechanization was displacing from the land, and recent immigrants from Italy, Greece, and Slavic lands. His employment practices were quite enlightened for the time: he hired the handicapped as well as thousands of African Americans and provided opportunities for ex-convicts to make a new start. His embrace of equal-opportunity practice made all the more puzzling, to friends and biographers alike, the virulent anti-Semitism of his writings and public pronouncements, anti-Semitism that he seemed later to regret.

Ford's engineers did what they could to make workers comfortable at their stations. But the drive to reduce production costs dictated faster machinery and speed-ups on the line. The tedium of assembly line repetition coupled with fast-paced production frequently produced synergies of stress and boredom that reached intolerable levels. These pressures, combined with the loss of any sense of vocation that craftsmen once felt, manifested in absenteeism, labor unrest, and excessive levels of employee turnover. These are still difficult issues that remain with us even today.

Worker alienation notwithstanding, Ford's methods became widely emulated. Assembly lines spread to companies supplying Ford with components to other automobile manufacturers and to producers of an increasing range of other products. Ford made no secret of his methods and arranged tours of his plants for others to see. This accelerated the pace at which his approach to producing a single, standardized product at high volume and low cost went from Ford to the plants of competing industrialists.

But time is no friend of engineering innovation—no friend if engineering loses flexibility, becomes entrenched, and is impervious to technological paradigm shifts. Ford fell into these traps. Over the nearly two decades of Model T production, engineers made design refinements to shave production costs, but the improvements became increasingly minuscule as time went on. More fundamental redesign to accommodate technological advances or evolving customer tastes were not in the offing. Typical of this rigidity was the comment Ford reportedly made, "You can have it in any color, as long as it is black." Likewise, as the company's engineers refined the production machinery for increased speed and reduced operating expense, they sacrificed flexibility. They optimized each single-purpose machine tool for producing a single Model T part, but reprogramming the machine to produce significantly different parts became all but impossible. Ford's fixation on producing a single standardized automobile left him vulnerable.

Alfred P. Sloan, leader of the newly formed General Motors, was not so entrenched in his thinking. He was free to transform Ford's production methods to his own ends. More important, he was attuned to technological developments capable of improving automotive design a great deal. At General Motors they followed changing consumer taste, the desire to go beyond the Model T's utilitarian ethos in styling, features, and performance, even if their cars were somewhat pricier.

Ford's competitors melded marketing with automotive engineering. And whereas Ford had been focusing on what we would now call "design for manufacture," his competitors led in the growing field of industrial design. Industrial designers were artists who applied their talents to shaping mass-produced products to be aesthetically pleasing and easy to use, paralleling much of what an architect would do in the design of a building. They designed automobiles with more streamlined styling and worked with engineers to translate evolving consumer tastes and desires into hardware. With employers and colleagues attuned to consumer preferences, engineers smoothed the ride with better suspension systems, increased the engine power, improved the steering and drive train, and incorporated electrical starters, windshield wipers, and other amenities.

Sloan was ushering in a more dynamic era, one in which automobile marketing was finely attuned to sleeker styling, and the engineering improvements were incorporated into a progression of new models. He also trumped Ford in manufacturing. Learning from Ford's manufacturing techniques and their vulnerabilities, General Motors devised more flexible systems for automobile manufacture. Machine tools were less specialized than Ford's: they were repro-grammable to fabricate parts for new designs with relatively little effort. Likewise, General Motors' engineers could more easily recon-figure their assembly lines to take on production of new models. This, coupled with careful planning, cut plant shutdown times needed for retooling. General Motors could change over manufac-turing facilities quickly to bring out new models annually.

Competition from more advanced automobiles caused Model T sales to drop precipitously from over half of the market in the early 1920s to 15 percent in 1927. Meanwhile, the Chevrolet, General Motors' most economically priced car, overtook the Model T in sales, amounting by 1927 to more than a million per year. The Model T's situation was then untenable. The late-in-coming, low-level design changes to make it more competitive failed to stem the fall in sales, and production ended in May 1927.

His competitors' success convinced Ford to end his long procras-tination. In the summer of 1926, he countered, instructing his engi-neers to design the Model A, a much-improved automobile that cap-tured many of his competitors' advances. But the new model brought to light larger problems. His production facilities were so rigidly specialized to the Model T that little could be salvaged. His engineers had to scrap many of the machine tools and design, build, or purchase new ones. Likewise, the Model A design forced them to reorganize the rigidly laid-out assembly processes. The necessary shutdown, retooling, and start-up of Ford's factories required an inordinate length of time. Although Ford predicted a smooth transi-tion, more than six months elapsed between the production of the last Model T and the assembly of sizable numbers of Model As.

Eventually, Ford's engineers overcame the changeover obstacles, and production rates increased rapidly in 1928. The Model A's mod-ernized design and features were popular with the public, but never did it dominate the market as the Model T had. General Motors, Chrysler, and other manufacturers had gained footholds though innovative automotive design but equally through the development of flexible mass-production methods. That flexibility continued to grow. We see it today in production methods that allow not only streamlined retooling to quickly change from an old model to a new—or even from one product to another—but also in the ability to produce on a single line products with flexible combinations of features and performance specifications. In contrast to Ford's day, flexible manufacturing is now ubiquitous. We see it in the plant where autos with different combinations of engine size, color, upholstering, and accessories emerge from a single assembly line. It appears even more prominently in built-to-order assembly, in computers, for example, where each buyer has specified the size and performance of the microprocessor, memory, and disk drive. The customized computer takes shape as its chassis moves down the assembly line.

Exacting precision, high degrees of mechanization and automation, and carefully orchestrated production characterize flexible mass production today. New developments in computer-controlled machinery, just-in-time parts delivery, statistical process control, and Internet connections between production facilities and suppliers add to the continuing quest for manufacturing productivity. Designers combine artistic talent and engineering prowess with scientific understanding to conceive what is to be built. The products and systems that come from their drawing boards—or more accurately their CAD systems—are truly marvels of our time. But if the methods for turning their creations into reality were not carefully thought out—if they were constructed one at a time, without the benefit of modern manufacturing methods—their costs would be prohibitive for all but a very wealthy few. Now, as in decades past, the production engineers, those whose focus is not on what to build but on how to build it, are pivotal to producing the wide variety of products that we enjoy at affordable prices. Rarely seen by the public, the ingenious machinery they design and the complex manufacturing organizations they create underlie our economic wherewithal.

❀   ❀   ❀

No one can deny the explosive increases in living standards that industrialization has brought. But momentous societal change accompanied industrialization. This change was often painful, particularly to workers displaced from craft occupations and to those subject to the abhorrent working conditions that all too frequently characterized the early phases of industrialization. The conditions in eighteenth-century England, particularly in the center for the mechanization of textile production at Birmingham, were among the worst. They captured the attention of literary figures and political philosophers alike. William Blake's poetry lamented England's "dark Satanic mills," and later, in *Hard Times*, Charles Dickens wrote of workers' deprivation in his fictional Coketown. Intellectuals as diverse as Thomas Jefferson and Karl Marx searched for social alternatives for obtaining the fruits of industrialization without creating the conditions found in England's factory towns. Neither Jefferson's vision of an American agrarian democracy nor the communist utopia sought by Marx stood the test of time. Blunting the impact of industrialization's darker ramifications was to be more an evolutionary than a revolutionary undertaking.

The twelve-hour workdays, child labor, dangerous machinery, and hazardous environments suffered by early wage laborers have all but vanished in Western democracies. Social reforms, union advocacy, engineering advances, and enlightened business practices all contributed to the long struggle to improve the lives of those who toil in the factories of industrial societies. The abhorrent conditions that so often characterized early factory life have been replaced by eight-hour days, time-and-a-half pay for overtime, and paid holidays. Mandatory safety standards for machinery and occupational health regulations guard against hazards, and workers' compensation laws provide redress when injuries do occur.

Nevertheless, the monotony of production line work and the occupational obsolescence that frequently accompanies industrial advance are matters warranting continuing concern. They became real for me one summer decades ago in a Chicago plastic factory, when I ran an injection molding machine to earn money for college. That molding machine, submerged in my memory, resurfaces whenever I visit an assembly line, a parts fabrication facility, or other sites of mass production. The machine's cycle time was about the same as the auto assembly plant I recently toured—one minute. In that minute I was simply to remove four plastic parts from the mold, check them for defects and push a safety clear to allow the next cycle to commence. These tasks didn't require a full minute, leaving a brief respite within each cycle. The problem was that the respite was too brief for me to leave the machine, to read or to engage in any other preoccupation. Thus throughout the day my mind wandered: I daydreamed and counted the minutes to quitting time, to lunch, and even to the next coffee break, anything that would free me from the monotony of the machine's cycle. The weekend and the return to school in the fall seemed like far-off dreams. The cycle's interminable repetition became so ingrained that I would wake in the middle of the night, unconsciously aware in my slumber of the injection molder's relentless rhythm of bangs, hisses, and thumps.

I often wonder what happened to those molding machines— and to my coworkers, not fortunate enough to attend college. The machines, I'm sure, were retired long ago, most likely replaced by a succession of more advanced machines, each producing parts more rapidly and more precisely and at lower cost than the last. It also seems likely that by now the machines are fully automatic, requiring no operator at all, or that production has been moved offshore to a developing country. I wonder whether my coworkers were able to upgrade their skills to the higher levels required for gainful employment in more modern production facilities. If not, their negative reactions toward technological advance would be understandable. They would be joining a long Luddite tradition of workers who resisted—sometimes violently—industrial advance at the expense of their livelihoods.

Engineering occupies a key role in determining not only what is to be made but also how it is to be made. It is this latter task, the design of efficient production processes, that has been central to driving costs down and creating ever-rising standards of living. The historical transformation from craft to modern mass-production methods created profits for investors, a profusion of affordable goods for the consumer, and increased wages for workers. Issues concerning how the fruits of industrial progress are to be divided among these groups go far beyond the technical challenges with which engineers most routinely deal. They are rightly issues for the broadest public debate and subject to the citizenry's political decision-making processes. But job safety, monotony, technological obsolescence, repetitive-stress ailments, and other day-to-day concerns of production workers interact strongly with the economics of production to affect the practice of industrial engineering—the practice of determining how things are to be made.

Eight

Fascinating Bedfellows

My notebook computer is becoming an almost constant companion. I rely on it not only for keeping books, doing taxes, preparing manuscripts, and putting presentations together but also for performing engineering calculations of a complexity that would have challenged even the most powerful supercomputers of only a couple of decades ago. With wireless communication to high-speed networks I'm in instant contact with friends and colleagues continents away, exchanging words and pictures with incredible ease. What is more, the Internet brings fingertip-access online banking, late-breaking news, satellite photos, market reports, and troves of other valuable information while also offering ease in comparison shopping for anything from mattresses to mortgages.

Add to computers the cell phones, satellite television, global positioning devices, medical imaging technology, and the wealth of other advances that we are currently witnessing, and the pervasive impact of technology—particularly information technology—on our lives comes more clearly into focus. None of these advances would have been possible without a deep knowledge derived from basic research in the physical sciences, in condensed-matter physics, quantum electronics, crystal structure, polymer chemistry, and a host of related disciplines. From such research the integrated circuits, fiber optics, and numerous related devices that stand at the base of today's information technology emerged. If we seek to understand how this wellspring of discovery is engineered into the commercial products that serve us daily, we will find that the process differs fundamentally from the engineering of earlier times, but resemblances still remain.

Since the seventeenth century, when science emerged from its earlier form to what we know today, the give and take between science and engineering has created a pair of fascinating bedfellows. Until the nineteenth century, scientific discoveries contributed little to the creation of new technology. Rather, it was the use of science's methods to analyze and improve existing machines and structures that was its primary contribution to the engineering endeavor. Arguably, these engineering investigations contributed more to the advancement of science than scientific discovery gave to engineering. Recall that in seventeenth-century Italy, Galileo parlayed his study of an engineering problem, the sagging of scaffolding in the Venetian shipyards, to the beginnings of the science of the strength of materials. And later, in eighteenth-century England, Benjamin Robins, in attempting to improve the predictions of cannonball trajectories, made the first observations of the sound barrier. At the turn of the nineteenth century, one of the more celebrated advances in scientific understanding began to unfold, with its roots in attempts to derive design rules that would predict the behavior of new steam engine designs.

Until the Boulton-Watt patents lapsed in 1800, Boulton and Watt's fear of boiler explosions caused them to insist that steam engines operate at atmospheric pressure. But with the monopoly removed, more adventurous engineers began to design and build engines that employed steam at higher pressures. By increasing the boiler pressure, they reduced the size of the engines while at the same time increasing their power. At high enough pressure, it made sense to eliminate the expensive condenser and exhaust the steam directly to the atmosphere. This, coupled with higher pressures, cut an engine's dimensions, weight, and cost and made steam viable for powering ships and railroad locomotives.

But a puzzle was in predicting the duty—or efficiency, as we now call it—of these higher-pressure engines. Duty—the amount of mechanical work obtained per bushel of coal burned—increased with each new engine design. Such increases were attributable in part to improved mechanical construction, decreased friction, and reduced heat losses. Steam pressure, however, seemed to be the central determinant of duty, and engineers sought to understand the effects that further pressure increases would have on an engine's economy. Answers to the pressure question would go far toward providing design rules, allowing designers to predict the power and efficiency of new engines before they built them. This, and the search for answers to related engineering problems, set a scientific agenda that contributed greatly to one of the nineteenth century's most fundamental advances in the understanding of the physical world.

The first breakthrough came nearly a quarter of a century after higher-pressure engines began to appear. Young French engineer Sadi Carnot published *Reflections on the Motive Power of Fire* in 1824, which showed that no heat engine could function without heat flowing from a hot body (the boiler) to a cold body (the condenser or the atmosphere). Hypothesizing an ideal engine, in what now is called a Carnot cycle, he proved that its maximum efficiency is determined by the difference between the temperature at which heat enters the engine and the temperature at which it is discharged to the environment. This provided new insight; attention should be focused on increasing temperature rather than pressure. It also provided a standard against which engineers could measure the efficiencies of operating engines, knowing that heat dissipation, friction, and other losses would prevent them from reaching Carnot's theoretical limit.

Practical considerations, the relationships between temperature, pressure, heating, and pumping in a family-owned brewery, served as motivation for young Englishman James Prescott Joule to study the relationship of heat to mechanical work. His experimentation grew to encompass the heating of wires caused by the flow of electricity, the transformation of chemical reactions to electricity in batteries, and other electrical and mechanical phenomena as well. Following a series of convincing experiments in 1843, Joule set forth the equivalence between heat and work, and a few years later he included electricity and chemical reactions along with heat and mechanical work in expounding his formulation of the law that energy must be conserved.

The scientific effects stemming from the study of heat and work multiplied. The caloric theory, in which heat was thought to be a weightless fluid that flowed from substance to substance, was abandoned, and Joule's dynamical theory, in which heat resulted from molecular motion, permanently took its place. R. J. E. Clausius and William Thomson (later to become Lord Kelvin) synthesized the work of Carnot and Joule to form the science of thermodynamics, the science setting out the relationships between heat, work, and energy. Technological results accrued as well. W. J. M. Rankin, an eminent engineer with close associations with the Scottish shipbuilding industry, was a pioneer in relating thermodynamics to improving steam engine design. He also examined the possibility of engines based on fluids other than steam. Air engines, for example, might achieve higher efficiencies without the safety problems of bursting boilers that high-pressure steam presented. And the air engine, in fact, did come into its own later in the century, in the form of the internal combustion engines of Nicolaus Otto and Rudolf Diesel.

The development of thermodynamics and the understanding of heat and of energy conservation is representative of sequences of events in which technological problems often lead to deeper insights in physics and chemistry. Investigations of engines, structures, or other devices may be motivated initially by the desire to improve designs, predict performance, or eliminate weaknesses. But such efforts often reveal heretofore unstudied phenomena, uncover effects that point to holes in scientific understanding, and lead to new avenues of scientific research. The results frequently do much to advance our understanding of nature in addition to providing the remedies to the practical problems that the engineers had set out to find.

Rich scientific dividends have continued to accrue through the efforts of engineers to gain understanding of their technological creations. Thus it was, for example, that Thomas Edison's 1883 study of the undesirable blackening of lamps raised questions instead of answering them. Under certain conditions of vacuum and voltage he observed a blue glow between two legs of the lamp's filament. The unexplained phenomenon became known as the Edison Effect. It underwent subsequent scrutiny that led in 1897 to a major breakthrough in fundamental physics, when J. J. Thomson published his theory of the electron. Likewise, the 1930s work at Bell Telephone Laboratories was focused on reducing static interference in overseas radiotelephone service, which had recently been established. Karl Jansky identified thunderstorms as one source of the static, but another, which he described as a steady hiss, turned out to originate from outer space. Study of the hiss led to the establishment of the field of radio astronomy and to astronomical discoveries made possible only by radio frequency observations. Similarly, attempts to understand why shortwave radios transmitted over much longer distances than expected led to the discovery of the ionosphere and also to the study of its properties using reflected radio frequency waves. Other examples abound. Many investigations flowing from the design and development problems of engineering endeavors have led to fundamental scientific advances.

While such advances continued to be made, a profound change in the interactions between science and engineering began to gain momentum early in the nineteenth century. Through the rapid sequence of discoveries in electricity and magnetism and the placement of chemistry on a quantitative footing, as well as other strides, nineteenth-century science blossomed in its ability to explain the physical world. Since that time, not just scientific methods for attacking problems but *the knowledge gained from basic science itself* became immensely important to engineering. Science's discoveries became paramount as a source of revolutionary new technologies.

With these scientific advances, technological innovation moved away from the insights of unschooled inventors and the visual world of beams, columns, gears, levers, cranks, and cams. Technological advances increasingly stemmed from those conversant in the sciences, the invisible world of molecules and atoms, electron flows, electromagnetic induction, and chemical bonding. Phenomena discovered through scientific investigation increasingly became the points of departure for technological innovation. Products and systems engineered to derive economic value increasingly were embodiments of abstract principles of physics or chemistry. The scientific study of electrical phenomena and their relationship to the development of telegraph systems offers an early example of these synergies between science and engineering.

Prior to the seventeenth century, common experiences with the emission of sparks, the crackling noises that sometimes accompany the rustling of woolen fabrics, and other electrical phenomena were largely ignored or viewed with superstition. In 1600, however, English physician William Gilbert published his book *De Magnete* (On the Magnet). The book was seminal in documenting the new approach of experimental science as he applied it to the study of magnetic phenomena. In it, he distinguished between electrical and magnetic attraction. His work thus marked the beginning of the scientific study of electrical phenomena. Scientists of the eighteenth century brought electrical phenomena, particularly those related to static electricity, under increased scrutiny. They developed the Leyden jar and other apparatus for storing increased amounts of electrical charge and studied its discharge. Benjamin Franklin's famed kite experiments in 1752 punctuated the century by identifying lightning as an electrical discharge. And his invention of the lighting rod marked the first technological advance stemming from the study of electrical discharges.

Leyden jars used in groups (then referred to as batteries) supplied larger electrical discharges, powerful enough to be detected over longer distances. In England, William Watson attempted to measure the speed at which an electrical charge traveled along a wire. He first attempted to measure the time it took electricity to transverse a wire stretched across the Thames, at a distance of twelve hundred feet, and later over a four-mile distance. He concluded from these experiments that the velocity of electrical conduction was "nearly instantaneous."

The nearly instantaneous propagation of electrical signals along long wires fired imaginations for possible means of rapidly communicating information over long distances. Such a technology would find much use, particularly by the military. At that time military engineers pursued elaborate methods for visual signaling. The most advanced of these was the invention of engineer Claude Chappe, which came into being in Napoleonic France. It consisted of a series of buildings positioned at intervals of several miles. Atop each building, as depicted in figure 42, an apparatus consisting of two arms, each with an extension, was located. The arms and their extensions were rotated to different angles, with each unique combination of positions corresponding to a letter or a number. Thus information could be transmitted visually by what is known as a semaphore code. But such systems left much to be desired. Transmission rates were slow, and operation was limited to clear weather and daylight hours.

More scientific study and discovery were essential before the lightning speed with which electricity traveled though wires could lead to a viable alternative to visual signaling. The first breakthrough came with Alessandro Volta's invention of the battery at the turn of the nineteenth century. By placing copper and zinc plates in a row of cups filled with brine and hooking them in a series, as shown in figure 43, he created a continuous flow of electricity through a wire loop. In 1820, experiments of Hans Christian Ørsted related electricity and magnetism and were also prologue to the discovery of the electromagnet. He found that a compass needle placed in the vicinity of a wire through which he passed a current deflected perpendicular to the wire. Then, in 1831, Michael Faraday discovered electromagnetic induction. He created a current in a coil of wire by plunging a magnet through its center. Soon after, Georg Ohm and André-Marie Ampère placed the relationships among potential (voltage), current, and resistance on a quantitative footing. William Sturgeon found that an iron rod wrapped with a coil of wire carrying an electrical current created an artificial magnet. US physicist Joseph Henry analyzed this phenomenon and showed that with a large number of

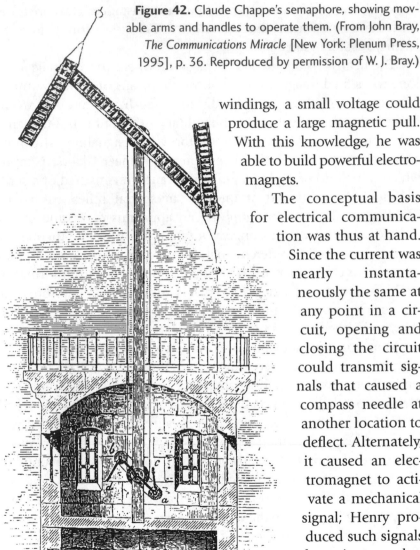

Figure 42. Claude Chappe's semaphore, showing movable arms and handles to operate them. (From John Bray, *The Communications Miracle* [New York: Plenum Press, 1995], p. 36. Reproduced by permission of W. J. Bray.)

windings, a small voltage could produce a large magnetic pull. With this knowledge, he was able to build powerful electromagnets.

The conceptual basis for electrical communication was thus at hand. Since the current was nearly instantaneously the same at any point in a circuit, opening and closing the circuit could transmit signals that caused a compass needle at another location to deflect. Alternately, it caused an electromagnet to activate a mechanical signal; Henry produced such signals by using an electromagnet to attract an iron armature and thus ring a bell. But while the basic science was in place, a great deal of what we would now call engineering development remained, beginning with a workable conceptual design, before the growing understanding of electricity could be deployed as a useful system for communication.

Figure 43. Alessandro Volta's series of cups filled with acid, forming the first electric battery: pairs of zinc and silver disks in a series of cups filled with acid. (From *Transactions of the Royal Society*, June 1800. Smithsonian Institution neg. 80-18643.)

During the first half of the nineteenth century, several inventors attempted to build systems for communication from the newfound understanding of electrical phenomena. The system devised by William Alexander, demonstrated in Edinburgh, brings to light some of the engineering challenges that a successful system must overcome. His apparatus, shown in figure 44, consisted of thirty wires and a copper rod, which he used as a common return. He associated a letter or punctuation mark with each of the wires. When he connected a battery to one of the wires, it closed the circuit and caused a magnetic needle to move, uncovering the associated letter on the panel. Thus by connecting and disconnecting circuits in proper sequence, he could transmit a sequence of letters to form a word.

Alexander's prototypical system highlights two major barriers between the concept of electrical telegraphy and its implementation as a commercially feasible system. First, the system was too complex. The thirty wires required for transmitting signals would lead to considerable difficulties in installation and maintenance, multiplying the cost far beyond a system using one or a few wires. Moreover, the rate of message transmission would be unacceptably slow. Second, this was a laboratory demonstration transmitting information over only five feet within a room! The combinations of coil and batteries used were neither powerful nor robust enough for transmission over long distances outdoors. Even a system that would function with a greatly reduced number of wires still required apparatus that would transmit signals reliably over hundreds of miles and under extreme weather conditions.

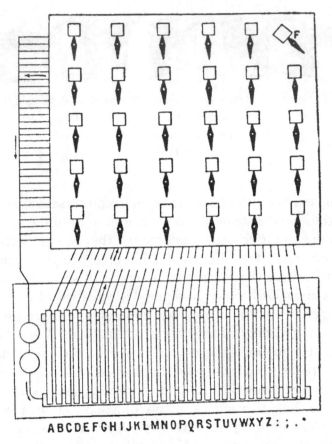

ABCDEFGHIJKLMNOPQRSTUVWXYZ:;.•

Figure 44. William Alexander's telegraph, in which a moving magnetic needle indicates the letter *F*. (From *La Lumière Electrique* 8 [1883].)

With the passage of time the needed technologies finally emerged. Higher voltage and longer-lasting batteries came into being. Henry devised more powerful and efficient electromagnets. Investigators gained a better understanding of the electrical resistance and insulation of construction materials and devised more suitable combinations of batteries and magnets (what we now call impedance matching) to keep signals transmitting and not reflecting back along the lines.

In the 1830s the resulting improvements in transmission capability and schemes to reduce the number of wires led to demonstrations of several early telegraph systems over increasing distances in Germany, Russia, and England. The most successful of these was that developed by English physicist Charles Wheatstone and his entrepreneurial collaborator, William Cooke. Using Ørsted's discovery directly, they employed compass needle deflections for converting currents back to letters and numbers. They reduced the number of wires needed to five by using the scheme shown in figure 45, which allows two compass needle deflections to point to a single letter, in this case V. Later simplifications made their system suitable for railroad signaling, where it found its first widespread use. Eventually, in 1848, they configured their apparatus to require only a single wire and needle, and the resulting system became the prominent means of telegraphy in Britain until the beginning of the twentieth century.

On the other side of the Atlantic, Samuel F. B. Morse was the first to conquer the multiplicity of engineering problems and assemble a functional system. Morse, an established painter, had longstanding interest in the electrical sciences. His idea for the telegraph came in 1832 while returning from an extended trip to Europe. On that voyage a conversation with a Boston chemist, Dr. C. T. Jackson, concerning the possibilities of electrical communication triggered his imagination.

Drawing in a small notebook, much as he might conceptualize a large painting with thumbnail sketches, Morse sketched a series of diagrams that in fact were variations on a conceptual design for a telegraph system. Central to his vision was what we would now call the software—what became the Morse code. Specifying each letter and number with a short series of dots and dashes, he could in principle transmit all needed information sequentially over a single wire. But many important questions remained to be addressed. At one end of the line he needed a means to rapidly open and close the circuit, in order to convert the dots and dashes to electrical signals. At the other he required a suitable device for converting those signals to markings or sounds from which the receptor could reconstruct the message. His system would require more operator training than the multiple-wire needle systems being advocated in Europe. But the

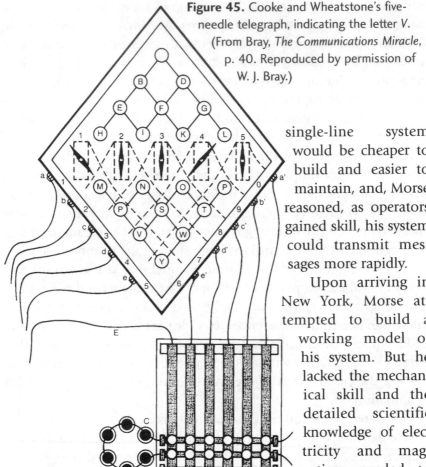

Figure 45. Cooke and Wheatstone's five-needle telegraph, indicating the letter *V*. (From Bray, *The Communications Miracle*, p. 40. Reproduced by permission of W. J. Bray.)

single-line system would be cheaper to build and easier to maintain, and, Morse reasoned, as operators gained skill, his system could transmit messages more rapidly.

Upon arriving in New York, Morse attempted to build a working model of his system. But he lacked the mechanical skill and the detailed scientific knowledge of electricity and magnetism needed to succeed. However, he secured a position as professor of the literature of the arts of design at New York University, which afforded him the time to pursue telegraphy. Equally important, Morse persuaded a fellow faculty member, scientist Leonard G. Gale, to collaborate in his venture. Gale had once worked with Joseph Henry and remained conversant with his research. Gale understood the need for more powerful batteries to overcome line resistance, and he knew how to strengthen an electromagnet's power by increasing the number of windings.

Already in 1830 Henry had shown that he could operate an electromagnet through a thousand feet of wire; he demonstrated how with sufficient battery power an electromagnet, located a mile away, could operate a clapper and strike a bell. Henry also invented the relay, a device that employed an electromagnet powered by the current in one circuit to open and close a second circuit. He successfully demonstrated the relay in 1836–37 by adding it to a circuit that he had set up two years earlier for class demonstrations at Princeton University. Such relays, placed periodically along a line, detected a faint signal, amplified it, and retransmitted it along the next section of line. With such relays, long distances could be divided into circuits of reasonable size, each equipped with its own set of batteries and linked to neighboring circuits by relays, or "repeaters," as they came to be known.

Henry realized that he had the components for a complete system of telegraphy. But he concentrated on scientific pursuits and made no attempt to patent his devices or to engineer a commercial system. Implementation of a commercial system, however, was the goal that motivated Morse. Gale's familiarity with the work of Henry and other physicists enabled Morse to combine more powerful batteries, electromagnets, and relays to increase transmission distances and reliability dramatically. A second partner joined Morse. Alfred Vail, a recent NYU graduate, brought much-needed skill in the design of mechanical hardware to the project. Additionally, his father, the owner of a successful iron works, agreed to finance construction of the equipment needed to test the prototype telegraph.

The hardware that Morse, Gale, and Vail settled upon combined a simple key for opening and closing the circuit and a recorder in which an electromagnet triggered a pencil to mark dots and dashes on a moving strip of paper, as shown in figure 46. The three partners perfected an apparatus that was more rugged and cheaper to construct than the European systems, and by 1837 they were able to telegraph information for ten miles using the Morse code dictionary. They publicly demonstrated the new system in 1838. Morse patented his telegraph system in 1840, and after much lobbying he obtained a congressional appropriation in 1843 to build a line between Washington, DC, and Baltimore. He first attempted to build an underground line, but that failed. Then at the suggestion of a young engineer, Ezra Cornell, success came by stringing the telegraph wire between wooden poles that were insulated from the wire with glass.

In 1844, following completion of the Washington-Baltimore line, Morse sent the famous message, "What hath God wrought?" He organized the Magnetic Telegraph Company in 1845 and built a profitable line between Washington and New York. Thereafter the telegraph spread rapidly to Boston, Pittsburgh, and Cincinnati, and by 1848 the telegraph network included every state east of the Mississippi except Florida. A transcontinental line to California went into operation in 1861, and by 1865, at the end of the Civil War, the United States telegraph network amounted to two hundred thousand miles of line.

The telegraph network eventually extended not only countrywide but worldwide. Linkages between the telegraph systems of European countries began late in the 1840s. Meanwhile, British engineers tackled the considerable technical challenges of laying undersea and eventually transoceanic cable. The system fathered by Cooke and Wheatstone in Great Britain first was connected to continental

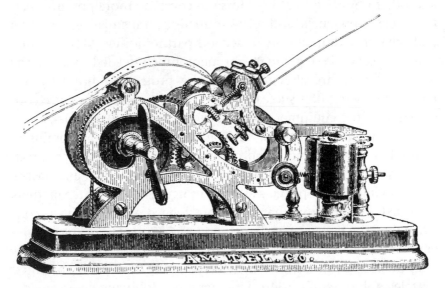

Figure 46. Morse telegraph for recording dots and dashes on moving tape.
(From George B. Prescott, *Electricity and the Electric Telegraph*, 6th ed.
[New York: Appleton, 1885].)

Europe in 1851 and to Ireland in 1853. The first transatlantic cable between Europe and America began operation in 1858 but failed shortly thereafter. It took Isambard Brunel's ship the *Great Eastern* to lay the cable that provided the permanent link between the two continents. With great fanfare, it went into operation in 1865.

As the telegraph system spread around the world, its capacity to carry more information on each line also advanced rapidly. For just as the Morse code allowed multiwired concepts to be replaced by a single-wire telegraph system, increased understanding of how electrical signals propagated through circuitry led to multiplexing. This allows transmission of several messages simultaneously over a single line. With increased efficiency, the telegraph's uses multiplied and had reverberations for government, industry, and commerce. News now traveled instantaneously from one part of the world to another, giving rise to daily newspapers. The pace of business picked up as international finance and commerce accelerated; markets for agriculture and industrial goods also became increasingly international in scope.

The telegraph's development highlights the growing linkages between scientific investigation and industrial innovation as the nineteenth century progressed. Michael Faraday, Joseph Henry, James Clerk Maxwell, and others who brought the electromagnetic sciences into being often had neither the interest nor the entrepreneurial drive to transform their discoveries into commercially viable technologies. But neither could inventors or entrepreneurs, such as Morse or Cooke, rely solely on their inherent ingenuity to create electrical technologies. Rather, they also had to understand the underlying science or, if not, work very closely with those who did. Thus in the United States, Morse the innovator teamed with Gale the scientist, and in England, Cooke the entrepreneur collaborated with Wheatstone the physicist to successfully engineer electromagnets, circuitry, voltaic batteries, and other electrical devices into workable systems of communication. Likewise, young telegrapher Thomas Edison read the papers describing Michael Faraday's experiments in electromagnetism to determine how they might serve as the basis for useful inventions. And Alexander Graham Bell attended public lectures and read extensively to understand the physics that in time would lead from the harmonic telegraph with which he experimented to his invention of the telephone.

It wasn't only these luminary innovators for whom the understanding of the science of electricity and magnetism was essential. The much larger numbers of those who translated their inventions into operable commercial systems required understanding of the underlying science as well. The expansion of telegraph—and later of telephone—systems required increasingly sophisticated circuitry, better insulators, and new devices, which included stock tickers, multiplexers, printers, keyboard terminals, and more. The "electricians," as they were then called, who designed these systems had emerged from the machine shop culture where the equipment was built or from the telegraph offices from where it was operated. The apprenticeships that many had served, however, required augmentation by extensive reading or formal education, for these engineers dealt daily with measurements and calculations of line resistances, signal strengths, and transformer ratios that would have been incomprehensible to unschooled craft workers. The discipline of electrical engineering grew from this need for more formal knowledge of electrical theory and experimentation among those who designed and built the communication and power networks of the nineteenth century. A parallel if quite different sequence of events gave rise to the discipline of chemical engineering.

Technologies that had sustained civilization for millennia had chemical reactions at their base: the smelting of iron ore, the brewing of beer, the tanning of leather, the bleaching of textiles, and many others. But chemistry as a science lagged behind the advances in physics during the seventeenth and eighteenth centuries. Likewise, application of the experimental methods of science and the use of mathematical theory came later to chemical technologies than to the engineering sciences relevant to structures and machinery. The chemical reactions that underpinned common eighteenth-century technologies were not understood or, for that matter, even recognized. Rather, craft traditions and rules of thumb ruled the day.

To establish a science powerful enough to apply to technological problems, chemical practitioners first had to dispel its close association with the techniques employed unsuccessfully in alchemy—the attempt to turn less precious substances into gold. This they did, and chemical theory emerged from its primitive state as the eighteenth century came to an end. Chemistry made immense strides early in the nineteenth century through the culmination of earlier investigations in a viable theory of the atom devised by English chemist John Dalton. By the 1860s, the work of Russian chemist Dmitry I. Mendeleyev on the periodic behavior of atoms placed the periodic table on a solid footing and provided an organizing principle for the understanding of chemical reactions on the basis of atomic theory.

Chemistry didn't first enter the practice of engineering as a wellspring for newly discovered phenomena, leading in turn to technological applications. Like physics, more than a century earlier, it entered first as a means of analyzing and improving existing technology. Chemistry was first used as a tool for analyzing the compositions and improving the quality of engineering materials—the iron and cement used by engineers, and of the ingredients of brewing, bleaching, and other commercial processes—based on chemical reactions. As the nineteenth century progressed, chemists tested, analyzed, and quantified existing materials and chemical processes. They examined material compositions and assessed impurity levels. Even though the characteristics they studied were not observable to the naked eye, chemists' analysis provided insight as to how these traditional industrial processes worked and how improvements might be achieved.

The smelting of iron ore and the production of steel provide important examples of the power of chemical analysis. Chemists analyzing ores for iron content and for impurities found the strength and toughness of the metal to be highly sensitive to minute variations in the ore's chemical composition. They put their knowledge into practice by analyzing material compositions and controlling impurities at the successive steps of the metallurgical processing. The results afforded by such analysis were often dramatic. In railroad technology, for example, the life of rails increased from two years to ten during the second half of the nineteenth century, and the freight weight that they could support increased nearly tenfold. Testing laboratories employing substantial numbers of chemists became common in the metals industries; for detailed understanding of chemical compositions and processing led to tougher, more durable alloys—alloys that maintained their strength at higher temperatures and allowed more powerful engines and more reliable machinery to be built.

Similar strides accompanied the application of chemical analysis to concrete, which had long been used as a building material. Chemical analysis and control of the raw materials used in the manufacture of cement—the lime, alumina, silica, iron oxide, and impurities—led to improved material properties. Chemists tried mixing the constituents in varying proportions. From such experiments they designed special cements tailored for a wide range of end uses. Both successes and failures challenged the chemists as they sought to obtain a more fundamental understanding of concrete's material characteristics and hardening properties. As in the case of metallurgy, by understanding the chemistry behind the industrial practice, they were able to improve the character and consistency of concrete.

Nineteenth-century chemistry, however, was to become more than a tool for analyzing the composition of materials. It became a research endeavor, and exploration for new chemicals and materials would lead to new technological developments. Perhaps the seminal event in initiating this new dimension of chemistry took place in England. There in 1856 William Henry Perkins, a young student at the Royal College of Chemistry, worked under the tutelage of Wilhelm Hofmann, a famed German-born chemist. Perkins was attempting to synthesize antimalarial quinine. Quinine was produced naturally from the bark of the cinchona tree of the Far East Dutch colonies, but natural quinine was in short supply. Perkins's experiments with an impure coal tar derivative called aniline resulted in producing not quinine but a strange black substance. From it he extracted a compound that he found stained silk to a rich mauve-colored shade of purple.

His accidental discovery had great commercial potential. For whereas the industrial revolution had brought great advances in the mechanized production of textiles, fabric coloring remained primitive, still relying on the rather monotonous hues of natural dyes that had been in use for centuries. Perkins patented the compound, and with the financial backing of his father, a prosperous London contractor and builder, he set about to produce the dye that launched the organic chemical industry. By the end of 1857, commercial silk dyers were buying the first aniline dye, tyrian purple. Further experimentation allowed him to extend the use of his dyes to cotton, wool, and print cloths.

Perkins's success encouraged others to search for organic compounds that would dye fabrics with other colors. Emanuel Verguin, a French schoolteacher, discovered a red aniline dye that he called *fuchsia*, and it became highly fashionable in the court of Napoleon III, under the name *magenta*. Within five years, more than two dozen firms from England, France, Germany, and Austria were producing five different colors of great popularity: mauve, fuchsia, imperial purple, a blue, and a yellow.

The discovery of these organic dyes, however, remained largely a matter of trial and error, for the scientific understanding of coal tar chemistry was in a primitive state. And while the aniline dyes exhibited color beautifully, they ran and faded easily and thus could not supplant the older natural dyes. These drawbacks caused each new color to enjoy only fleeting popularity, and those selling them could expect high profitability for only a few years. To overcome these difficulties and systematically develop new and improved products, the dye industry needed a firmer base for understanding the relevant organic chemistry.

The dye industry's focus soon moved to Germany, drawn by the unparalleled system of scientific research and technical education that was developing there. German universities—including the university at Giessen, where Perkins's mentor Hofmann had been trained—were becoming the world's leading institutions of scientific learning. Equally important, the first of Germany's famous *technische Hochschulen* (roughly translated, technical universities) opened in Karlsruhe in 1825. Modeled after France's École Polytechnique, similar technical universities quickly sprang up across Germany, combining scientific study with technical training. The ties between academia and industry were stronger in Germany than anywhere else in the world. Industry supplied universities with chemicals, undertook tedious analysis of compounds, and sent their staff chemists on leave to work in university laboratories. Industry provided consulting fees for prominent faculty, paid royalties for newly invented compounds, and provided legal support in patent processing. Industry, in turn, secured the rights for the discoveries that key researchers made in academic laboratories. It also secured a stream of employees with excellent scientific training but who were also familiar with the challenges of industrial problems.

Fundamental understanding of coal tar chemistry paid off in 1868 in an event that was key to German industrial expansion. Two Berlin academics synthesized alizarin, a much more colorfast dye than aniline. Alizarin made the power of scientific investigation more obvious as a source of commercial products. It and the other organic chemicals later synthesized in Germany's academic-industrial complex played no small part in the establishment of large chemical plants along the Rhine and its tributaries.

The need to produce large quantities of sophisticated chemicals economically and with a minimum of impurities created challenges that went beyond research laboratories. Chemists first performed the synthesis and testing of new chemical products by using small quantities in benchtop experimentation. But once an industrial firm decided that the product was commercially valuable and established a market, it knew that large engineering tasks remained. The procedures for making the chemicals had to be scaled up to produce the new product economically in large quantities. That required merging chemistry with other forms of engineering. Chemical reactions depend on temperature, pressure, mixing, and more, and systems of tanks, pipes, pumps, valves, compressors, evaporators, and other equipment had to be designed to carry them out on massive scales. The resulting need to combine chemical knowledge with that of structures, machines, and control systems gave rise to what later became known as chemical engineering.

❀   ❀   ❀

As the nineteenth century progressed, the rapidly advancing sciences of electromagnetism, organic chemistry, and related areas transformed engineering as they fostered new inventions and innovations. The emergence of electrical and chemical engineering exemplified the transition, as engineers required deeper understanding of the mathematical theories and experimental procedures of science. For only then could they conceive, design, and build the products and processes whose operational principles could only be explained in the quantitative language of science. Electrical engineers needed to calculate battery strengths, numbers of electromagnet windings, currents, voltages, and a host of more subtle electromagnetic quantities in the design of each new telegraph line. Chemical engineers were faced with predicting how the reaction rates and yields of the process equipment they designed would vary with temperatures, flow rates, catalysts, and constituent concentrations. Advancing science affected the older, more established fields of civil and mechanical engineering as well. These disciplines transcended the craft traditions from which they originated and incorporated scientifically based determinations of stress, flow, temperature, power, and efficiency into the design and testing of structures and machinery.

The growing necessity for scientific knowledge in engineering practice brought widespread change to how engineers prepared for their work. Historically, candidates sought entry into the profession by serving extensive apprenticeships under experienced practitioners. Most frequently, they began their careers working in small groups on construction projects or in machine shops, learning the surveyor's or machinist's trade before mastering the technical and organizational skills needed for taking responsibility for engineering endeavors. The professional goal of the more ambitious engineers was to become prominent independent consultants, moving from project to project or to establish their own businesses, from which they would direct the design and construction of structures or machines.

More formal preparations originated with the establishment of the famed French engineering schools of the eighteenth century, most particularly the École Polytechnique. These served as models for the technical universities in Germany, the military academies and then the land-grant colleges in the United States, and other schools established across Europe and America. Such institutions exposed engineers-in-training to mathematical theory, engineering science, and technical drawing. Throughout the nineteenth century arguments raged within professional societies, universities, and places of employment concerning whether apprenticeship or academic training provided better preparation for engineering practice. By the turn of the twentieth century, however, the on-the-job apprenticeship, augmented by outside reading, technical courses at night, and other part-time education was largely supplanted by career preparation that began with engineering education at the university level.

The nineteenth century also saw the focus of engineering effort move toward the design and construction of technological systems of greater size and complexity. Beginning with the railroads and the telegraph networks, engineers increasingly dealt with water treatment and sewage disposal systems for mushrooming cities, electric power and telephone grids, and in more modern times continent-spanning highway networks, air traffic control systems, satellite communications, and the Internet. By the end of the nineteenth century, even single—sometimes seemingly simple—products often involved many engineers in the planning of complex manufacturing and supply systems necessary for mass production. In concert with such change came new institutional organization of engineering practice. Private practice and employment in small shops and businesses became less common than larger engineering efforts carried out under the auspices of sizable corporations and government agencies.

The nature of technological innovation—what we now call *research and development*—also became more highly organized during the course of the nineteenth century. In the United States, Thomas Alva Edison was pivotal. For when he established his laboratory at Menlo Park, New Jersey, in 1876, he moved technological innovation away from the domain of independent inventors and from isolated shops and laboratories into the realm of more highly coordinated industrial research and development. Edison conceived the laboratory—often referred to as the "invention factory"—to bring scientific knowledge to bear on industrial innovation. His aim was to combine under one roof the practical ingenuity of inventors with scientists' understanding of the underlying principles and theory.

The "Wizard of Menlo Park," as Edison later came to be known, assembled the latest scientific instruments, a chemistry laboratory, and an excellent technical library. He began with a staff of fifteen that later grew to seventy-five. He brought together university graduates educated in physics and chemistry, mechanical designers out of the shop tradition, machinists, instrument makers, glassblowers, and more. They collaborated in performing experiments, making measurements, and designing and constructing electrical apparatus. But Edison was not motivated by the pursuit of pure knowledge. He focused the laboratory's investigations with careful consideration for eventual marketability of the products they might produce. He knew that such a laboratory could not sustain itself for long if its products had no value in the commercial world, since the business interests who invested in it would soon withdraw their support.

Descriptions of the laboratory's work atmosphere bring to mind analogies with today's computer hackers who are working in Silicon Valley start-up companies. Edison encouraged an open environment with few work rules, which was a great surprise to those accustomed to the more staid discipline of the shops and businesses of the day. Those who ran the experiments that hopefully would lead to useful devices were called "muckers." They and their assistants were highly motivated and worked long but flexible hours, staying up all night or longer when needed, then collapsing to recuperate. Camaraderie was strong, with intermittent storytelling and horsing around to relieve the tension of long hours and exacting tasks. Edison sought to bring out the creative powers of the muckers, working closely with them, sharing in jokes and in all-night stints. But the "Old Man," as his employees knew him, did not countenance theoretical science for its own sake. Thus even graduates of the best-known German and US universities had to prove their practical worth as muckers in designing and building apparatus before he accepted them into the brotherhood of Menlo Park.

Development of a system of electric lighting soon became the central objective of the Menlo Park laboratory. And *system* was the key word, for Edison had not only to invent the electric lamp itself but an entire system—including centralized generation (the dynamo) and distribution (the wiring and metering) to deliver the electricity to the offices and homes where it was to be used. To accomplish this, of course, his efforts extended well beyond the research and development efforts at Menlo Park. He formed companies to market the lighting, manufacture the equipment, and operate the system that emerged from the laboratory's work. The efforts were successful; only five years after inventing the incandescent light bulb in 1877, Edison opened the Pearl Street Power Station and supplied lighting to the homes of that central district of New York City. The laboratory was then closed, but soon after Edison built a larger industrial laboratory in West Orange, New Jersey, which inherited many of the traditions and personnel of Menlo Park.

Menlo Park was indicative of the changing institutional environments in which technological innovation would increasingly take place as it relied more and more on the discoveries and methods of science. Interdisciplinary teams of salaried engineers and scientists increasingly displaced the inventor-entrepreneur as the originators of technological innovation, and extensive laboratories supplanted small shops as the venue of innovation. In the last decade of the nineteenth century, increasing numbers of research and development laboratories appeared in the United States and Europe. Most differed from Edison's invention factory, however, in that they were appendages of large corporations.

By the 1870s the larger German companies of the synthetic chemicals industry—BASF, Hoechst, and Bayer—began forming their own corporate research laboratories. They understood the much wider commercial possibilities that could grow out of the raw materials, intermediate products, and processes that they first employed in dyestuff manufacture. They hired many chemists to discover new chemical products and explore new manufacturing techniques. As industry branched out into broader areas of chemical products, its hiring became more interdisciplinary. In addition to chemists—both organic and inorganic—they employed biochemists, bacteriologists, and biologists. Their expansion created a growing demand for separate research divisions housed in suitable facilities. To this end the companies created well-equipped laboratories for research and development. They employed instrument makers, laboratory technicians, and other support personnel in those laboratories to relieve the research workers from the more routine chores. They also provided impressive libraries of books and technical journals within their facilities for the investigators' use. Investment in these laboratories and their successor institutions was rewarded with a continuing parade of new products of pharmaceuticals, fertilizers, pesticides, photographic film, and synthetic fabrics and employed an increasing number of polymer materials where wood, ceramics, or metals were once required.

In the United States more than thirty firms had established research and development organizations by the end of the nineteenth century. For the most part, their efforts tended to focus on making incremental improvements to existing products. In 1900, however, General Electric's famed theoretician and chief consulting engineer, Charles Proteus Steinmetz, convinced his company's management to establish a laboratory that would distinguish itself and become a model for industrial research and development in the twentieth century. The company drew Willis Whitney away from the faculty at MIT to direct the laboratory's efforts and hired as its staff both practiced engineers and scientists holding doctorates from prominent European and US universities. The laboratory's management freed the investigators from day-to-day troubleshooting assignments to perform more basic research and to parlay their scientific findings into innovative new products and processes. Success flowed from these efforts and resulted in the company's expansion from lighting into power production, communications, medical instrumentation, and other newly discovered uses of electricity.

In good measure, the laboratory's success rested on the ability to attract the best scientists and engineers away from competing career paths. For the scientist, laboratory employment meant divided effort between the pursuit of knowledge and participation in projects that used that knowledge to create commercially viable technology. The alternative most frequently was college teaching, but there the time afforded away from the blackboard to pursue research often was less than that of the industrial scientist. Moreover, neither the equipment nor the pay was as good. Thus industrial employment was attractive to some, even though they might have to delay publication in scientific journals—at least until company lawyers locked in patent applications for anything of commercial value—or in some cases forgo publication altogether.

In contrast, the engineers' gratification came from designing and building new products or processes and then observing the success of their creations in use. This they might accomplish either as corporate employees or as independent workers. Since patent rights invariably went to the corporation, corporate employees had no chance to make the fortunes of Morse, Bell, or other independent inventors. However, many engineers, particularly those most interested in the technical heart of innovation, were willing to forgo the small chance of becoming exceedingly wealthy. For with corporate employment they avoided the extreme financial risks of the entrepreneur as well as the preoccupations with marketing, finance, and litigation that would draw them away from the technical challenges they loved.

The processes of technological innovation changed from the time of the telegraph to Edison's inventions and on to the research and development performed at General Electric and the other corporate laboratories that followed. As engineering moved into the twentieth century, the impact of science grew stronger, and not only as a rational approach for predicting performance and testing the results of designers' ideas as they moved from concept to detailed design to prototype. The discoveries of science became a powerful wellspring of ideas for new design concepts and technological innovations. Science—as a source of technology—brought about a transformation in the nature of invention.

Since the Renaissance and before, engineers had created ingenious mechanical devices, even though they sometimes served no immediate need. Their technological exuberance overflowed, proposing solutions to problems that didn't exist. Many such devices are illustrated in seventeenth-century "theaters of machines," books filled with conceptions for every imaginable purpose. But far more frequently, inventors focused their efforts on solving pressing practical problems: their challenge was to produce apparatus suitable to the task. In eighteenth-century England, for example, industrialists sought to replace the hand motions of spinning and weaving with mechanized methods for performing the same tasks, and Boulton and Watt targeted their first steam engines toward the growing need to pump water from the mines of Cornwall.

The nineteenth-century rise in the importance and power of science shifted that balance. The profusion of discoveries has since become a driving force for the creation of new technology. Each discovery now prompts dreams of new technological devices, even though it is often unclear what practical use they might serve. But more often than not, uses are eventually found, frequently important ones, and these we call "technology-push" innovations, as opposed to those prompted by specific need, or by "market pull." Microwaves—similar to radio waves, but with shorter wavelengths— offer a good example.

The scientific insights into the properties of microwaves gained from intense wartime radar research stand at the beginning of several such technology-push innovations. Both the microwave oven and the widespread use of lasers originated in that work. The development of these devices illustrates the synergetic effects of the intertwining of scientific discoveries and engineering development that spearheaded technological advance in the twentieth century.

Shortly after the end of World War II, Percy Spencer, an engineer with the Raytheon Corporation, was testing magnetrons, the vacuum tubes designed to produce microwaves. He found that the microwaves had melted a candy bar in his pocket. Fascinated by the cooked candy bar, he tried placing popcorn kernels near the magnetron, and sure enough in no time at all they popped and scattered about his lab. Joined by a colleague, the next day he tried another experiment that proved to be even more revealing. They exposed an egg to intense microwaves, and it also cooked, but not as they expected: the egg first began to quake and tremor, and then it exploded, scattering hot yolk about the room.

The cooking properties of microwaves were unique. The microwaves didn't penetrate metallic surfaces, but they did penetrate nonmetallic materials, such as candy bars, popcorn, or eggs, delivering heat directly to the interior. If the heated sample was not too large, the microwaves could heat it quite evenly as they were absorbed throughout the body's volume. This offered a possibility for a new and very different way for cooking. If food is placed in a traditional oven, the oven is set at a high temperature. The heat then diffuses slowly inward from the food's surface to the interior, cooking the material near the surface much more thoroughly than that near the center. Moreover, oven temperature and the slow diffusion of heat severely limit the cooking speed. In contrast, microwaves penetrate the interior instantaneously, depositing heat inside the food and cooking it more uniformly and more quickly.

These advantages of the microwave were obvious to the engineers at Raytheon. Spencer took the first step: he built a prototype microwave oven by feeding microwave energy into a metal box from which it could not escape. The high density of microwaves in the box caused food placed in it to heat rapidly and to cook in record time. But progressing from the heating capabilities of microwaves in this first prototype oven to a commercially viable product required a great deal of engineering, marketing, and business savvy.

Raytheon patented the process and tested the first ovens in a Boston restaurant. The first units were large—nearly six feet tall and weighing several hundred pounds—and cost thousands of dollars. Initial sales were disappointing. But with further engineering, sizes, weights, and costs came down. New industrial uses were found; microwaves also proved efficient for drying potato chips, roasting coffee beans, and precooking meats. Meanwhile, the ovens began to catch on with restaurants and vending companies, for their owners could keep food refrigerator-fresh and then cook it rapidly to order.

The larger potential market for microwave ovens, of course, was the home. But Raytheon was a high-tech defense contractor whose primary expertise was in engineering radar and other electronic equipment for ships, aircraft, and missiles. The corporation had experience neither in designing products for the home, nor in the distribution, marketing, and advertising needed to sell them to the homemaker. But by acquiring Amana Refrigeration, the Iowa maker of home appliances, Raytheon was able to combine its microwave expertise with the insights of Amana's industrial designers into the size, features, appearance, and other characteristics that an oven had to have to appeal to the homemaker. Equally important, Amana had the distribution channels to retail outlets and the advertising savvy to make the consumer understand the advantages of having a microwave oven in the kitchen.

Other problems had to be overcome to bring about the success of microwave ovens. Public acceptance required that safety concerns over unknown dangers from this mysterious new way of cooking be addressed; recipe books required modification; and homemakers had to experiment and become comfortable with the new way of cooking. Eventually, the nature of food preparation changed to take advantage of the speed and convenience of the new cooking method, and companies marketed foods—TV dinners, for example—specifically for microwave preparation. In many homes the new oven at first stood idle, except when used to boil water rapidly or for other peripheral uses. But as consumers came to appreciate the oven's speed, convenience, and energy efficiency, microwave cooking gained prominence, and consumers used their conventional ovens less and less. Thirty years after that first laboratory accident, microwave cooking had become firmly established, with the sales of microwave ovens surpassing those of gas ranges. Today the progression of microwave cooking from discovery to mature product is complete. The ovens are commonplace fixtures of kitchens, mass-produced in varieties of sizes, colors, and styles at minimal cost to fit into any household.

The laser, like the microwave oven, has its roots in research efforts prompted by wartime military needs, but its conceptualization followed a quite different route from the accidental recognition of microwave cooking. The laser is a device for emitting either pulses or a continuous beam of very pure light, very pure in the sense that it is a coherent beam of a single wavelength or color. Unlike other light sources, it can be made very intense while it is maintained in a sharply focused, pencil-like beam. Unlike the microwave oven, inventing the laser required deep scientific insight into atomic physics, for the laser's underlying physics is based on stimulated radiation emission, a phenomenon that can be understood only in terms of quantum transitions between atomic and molecular energy levels and the concomitant absorption and emission of radiation. The physical principles of stimulated radiation emission were first recognized in the 1930s, with none other than Albert Einstein playing a central role.

Not until after the storehouse of knowledge on the generation and detection of microwaves gained from World War II efforts became available, however, was the critical experimentation leading to the laser possible. Much of the technology needed to develop the laser came from microwave radar systems. Charles W. Townes, who was later to share a Nobel Prize for his laser work, had spent the war years dealing with the microwave science underlying radar and in particular gaining insight into the noise problem of microwave amplifiers. This work, in part, motivated the seminal contributions that he was to make.

Prior to the laser was the invention of the maser, which is similar to the laser in principle but operates with microwaves instead of visible light. In 1954 Townes and two young associates, James Gordon and Herbert Zeiger, successfully built an operating maser in Townes's laboratory at Columbia University. Then in 1958 Townes and Arthur L. Schawlow of Bell Laboratories generalized the concept from the maser success and published the classic paper that set down the principles for making lasers with electromagnetic radiation of any wavelength, ranging from microwaves through visible and ultraviolet light. That paper was also the basis for the Bell Laboratory patent on the laser. Within two years, Theodore H. Maiman of Hughes Research Laboratories built the first successful laser, emitting pulses of pure and intense light. Rapid developments then followed at IBM, Bell Labs, and other industrial laboratories. They produced lasers utilizing a variety of different materials, ranging from gases to semiconductors, to produce beams with colors ranging from the infrared to the ultraviolet.

The laser—like the maser before it—provided experimental confirmation of theory of stimulated emission. In its earliest days, however, the laser was little more than a scientific curiosity, and its creators were not yet sure what uses it might find. People kidded Townes and the other laser pioneers, characterizing the laser as a solution looking for a problem. And although the physicists who invented it thought that many uses would be found, it took sustained effort to connect these sources of very pure and intense light to practical application. And it took a great deal of engineering enterprise to bring those connections to fruition.

Not surprisingly, physicists found the earliest laser applications. They used the pure light to probe the structure of atoms and molecules, to induce chemical reactions in the laboratory, and to measure distance and time very accurately. Today virtually no laboratory, whether it is in physics, chemistry, biology, or engineering, is without its lasers. They are invaluable in studying fluid flow patterns, crack propagation, chemical reaction rates, and a host of other phenomena. But in the decades since their invention, lasers have moved outside the laboratory to find multiple commercial, industrial, medical, and military applications.

Perhaps the most visible laser use is in the ubiquitous bar code scanners used in grocery stores for checkout and inventory control. But there are far more uses hidden from our eyes. They read information to and from our computer discs and from our CDs and DVDs as well. Lasers are combined with photodetectors and fiber-optic cable to vastly expand telecommunications capabilities by replacing electrical with optical signal transmission. Manufacturing firms use lasers to replace more traditional methods for drilling, cutting and welding metals and other materials. Of the many uses of lasers, none is more gratifying than their use in medicine. They have been central to the development, for example, of minimally invasive surgery, in the development of endoscopic devices for treating ulcers, removing tumors, reattaching severed nerves and particularly for removing cataracts, repairing torn retinas, and dealing with other visual problems.

Many laser applications that are now commonplace were not obvious at the beginning. Their development required those involved to contemplate doing things in new and unfamiliar ways. To use lasers in surgery, for example, investigators first had to work out each new technique with intensive laboratory experimentation. They then faced proving a method's effectiveness and demonstrating its safety in studies on animals and cadavers before clinical trials on humans could begin. Once government agencies approved the new techniques, the medical innovators had to convince practicing surgeons, trained in the traditional surgical techniques, that the radically different approach would deliver superior results, justifying the expense of new equipment and the time and effort for retraining.

Microwave ovens, laser surgery, and a myriad of other technological advances have poured forth from the fruits of scientific discovery at an ever-accelerating pace. New materials and devices appear, often without obvious practical value. Scientists, engineers, and entrepreneurs alike struggle to envision the eventual uses that will make superconducting magnets, carbon nanotubes, micromechanical machines, and other recent products of our research laboratories valuable contributors to our well-being. The technical community exudes confidence that uses will be found for many if not all of these developments. This flow of scientific study and discovery that feeds the engineering development of new products and processes has established fundamental science as engineering's powerful benefactor. But even as in centuries past, the efforts of engineers to analyze and improve the technology with which they work also continue to provide insights that contribute to the advance of basic science.

Nine

Pushing the Envelope

On my way to Chicago's O'Hare International Airport a few years ago, I checked our tickets and found that my wife and I would be flying on a Boeing 777 to Tokyo. I had read much about the aircraft's development and introduction into service, but this would be our first opportunity to actually fly on it. The ease with which our carry-on luggage slid into the overhead bins gave witness to the improved design of the 777. As we took our seats, the entertainment system, the seat adjustments, and other visible signs attested to the advances that Boeing's industrial designers had made since the introduction of its earlier aircraft. Of course, the most significant engineering innovations were beyond our view: they included extensive replacement of metal components with stronger, lighter composite materials and the adoption of a digital fly-by-wire flight control system. The aircraft was impressive for its size, seating nearly four hundred passengers, but even more for its capability to cross the Pacific powered by just two engines, each more powerful than any previously built.

This plane is truly state-of-the-art, I thought as we sat on the runway awaiting our turn to take off. But as I looked out the window, I saw a 727 taxiing into its position ahead of us. It brought the point home to me that the 777 was indeed not revolutionary technology. It had evolved from the 727, designed some thirty years earlier, and from a whole series of airliners in between. With the exception of the 747, the 777 was larger and could fly more passengers than any of its predecessors. It could fly them for longer distances, more economically, reliably, and comfortably. Many of the components and subsystems brought together new concepts, hardware, and ways of doing things, but as a whole, the aircraft was not a radical departure from earlier designs. It wasn't the result of pushing a fundamentally new technology into the market but rather the result of market pull. Boeing responded to an opportunity, the need for an airliner with a specified range and seating capacity—one that capitalized on the technical advances achieved since the design of previous aircraft to better serve the airlines who would buy it and the passengers who would ride in it.

The aircraft's evolutionary design stands in sharp contrast to revolutionary design concepts—those seminal technological advances that allow new things to be accomplished—often things that hadn't even been contemplated. As noted in the last chapter, devices and processes arising from inventive minds and scientific discovery often first appear without an obvious use, and uses may take decades to develop. It was not clear to those who invented the laser, for example, that with time it would give rise to compact disc recording, bar code scanners, and laser surgery. Revolutionary design concepts often begin as solutions looking for problems, or, more precisely, as new technologies looking for markets to serve. In contrast, the 777 was a solution to a specific market need.

The 777 may not represent a technological advance of historical proportions. But that doesn't preclude it from being a magnificent piece of engineering. Indeed, most engineering is not centered about the introduction of a radical new technology. Most projects focus on improving products and evolving designs to meet changing and more demanding market conditions. They bring together state-of-the-art methods and technology to solve the problem at hand. They must meet their customers' demands without waiting for scientific discoveries yet to come or utilizing new phenomena yet to be understood. Yes, they may pull technological innovations out of research and development laboratories and into their designs, but those innovations must be sufficiently proven to limit the risks of inordinate delays and uncertain outcomes. Engineers must look for solutions now, solutions that they can secure with limited budgets and under the pressure of compressed deadlines. They design at the other end of the spectrum from technology push—they are responding to market pull.

As we taxied down the runway and lifted off, I contemplated the enormity of the engineering effort required to bring the 777 into existence. That effort began in understanding what the airlines' requirements were. It continued with a preliminary design, which the engineers then revised and refined until they produced the detailed sets of plans for the aircraft and for the tools and procedures for its manufacture. Throughout their endeavor, they computed, experimented, and tested first to predict and then to verify the behavior of their creation. The collective knowledge and experience that must be brought together to create such an aircraft is awesome. I recalled the title of engineer Walter Vincenti's insightful book, *What Engineers Know and How They Know It*.[1] For certainly the knowledge that went into the aircraft's creation represents the current state of the long historical quest for both analytic knowledge and tacit know-how: first to learn how fly at all, and then to bring flying to its present state of comfort, safety, and economy.

The 777 is among the largest and most complex of today's technological accomplishments. But the paradigms employed to create it share a common core with those central to the engineering of all material entities, whether they be bridges, buildings, automobiles, appliances, computers, or chemical plants. Engineers push the envelope of technology in their ongoing quest for a superior outcome: higher performance, greater dependability, and improved economy. And as they do, they must overcome challenges placed before them by conflicting requirements and limited knowledge, by restrained budgets and tight deadlines.

Each technological object must stem from a conceptual design, for the conceptual design defines that object. Engineers must understand the concept, the operating principles and general configuration of what they are to create, before they can proceed to design at a more detailed level. In normal or evolutionary design, the concepts are known to work; they have been hammered out by earlier generations of engineers. Though once revolutionary technological innovations—new inventions—now the architecture and operating principles are common knowledge deeply embedded in the minds of the design engineers.

The basis for the 777—and of all other airliners—dates back to George Cayley, who first conceptualized the fixed-wing aircraft at the turn of the nineteenth century. The fixed-wing concept differs distinctly in configuration and operating principle from the helicopter, the dirigible, and the wing-flapping concepts of flight. But within the conceptual umbrella of fixed-wing aircraft, engineering advances come with each new generation of airplanes. Design of any complex technological device is hierarchical, and new concepts and revolutionary design may be embodied at lower levels of that hierarchy in subsystems and components. Thus the 777's appearance is quite different from Cayley's conception or the Wright brothers' "flyer," even though they share the same operating principles and similar configurations of fuselage, wings, and tail.

The transformation from propeller-driven aircraft to jet engines offers a salient example. Jet engines constituted a revolutionary advance in propulsion systems, bringing about dramatic increases in aircraft speeds and altitudes. Propeller-driven aircraft had performance limits of about 350 to 400 miles per hour. For although the aircraft is then flying at only about half the speed of sound, the propeller tips are moving much faster. As those tips approach sonic speed, they experience increased drag, which makes the propeller's efficiency plummet. A propeller also loses efficiency in the thinner air at high altitude, sharply limiting the speeds and altitudes at which it can operate. In contrast, most jet-propelled aircraft cruise at speeds approaching the velocity of sound, and those designed for supersonic flight routinely pass through it. Moreover, a jet engine's efficiency increases at higher speeds and at higher altitudes, further expanding the envelope at which aircraft can operate. The 777 cruises at more than 80 percent of sonic velocity and at altitudes where no propeller-driven aircraft could venture. Moreover, as experience in jet engine use was fed back to improve the design, added advantages accrued. Its spinning turbines replaced the reciprocating motions of the piston engines, thereby reducing vibrations and providing a smoother ride. With fewer vibrations and fewer parts, jet engines also required less frequent maintenance.

Since the Wright brothers, there has been a continual stream of aeronautical innovations; metal took the place of wood for airframe structures, retractable landing gear replaced fixed wheels, single-wing aircraft replaced biplanes, then swept-back wings came into use, and the list could go on. But each of these relates back to the basic force diagram of drag and lift pictured on Cayley's coin (refer to fig. 19, p. 95), and the forces of thrust and weight. Within the Cayley-Wright lineage, aircraft built for similar purposes tend to look increasingly alike. The present-day Boeing airliners shown in figure 47 closely resemble not only each other but also earlier aircraft of both Boeing and its competitors dating back to the first jetliners of the 1960s. As technologies mature, they become more evolutionary and tailored to the needs of specific markets. But beneath the surface each of the pictured aircraft contained innovations not present in its forerunners, as engineers refined their designs to meet the demands of particular markets while at the same time incorporating technological advances that had not previously been available. The fly-by-wire navigation, composite materials, and improved engines of the 777 fall into this category.

Conceptual designs are only the beginning. Engineering is a quantitative profession, and before design concepts can be transformed into a detailed set of plans from which an aircraft can be produced, a bridge built, an automobile assembled, or a computer chip fabricated, specifications must be set. The needs of the patron, client, or customer for whom a project is carried out must be expressed unambiguously. Even then, meeting a multiplicity of performance specifi-

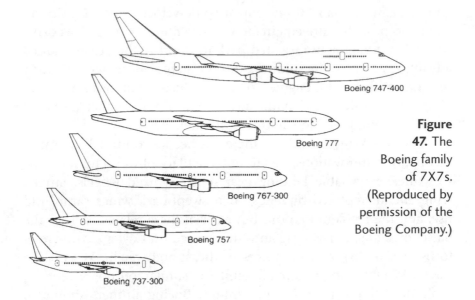

Boeing 747-400

Boeing 777

Boeing 767-300

Boeing 757

Boeing 737-300

Figure 47. The Boeing family of 7X7s. (Reproduced by permission of the Boeing Company.)

cations is likely to be one of the engineers' toughest challenges. But once the criteria are specified as numbers, the engineers at least know what they are shooting for. They know the targets they are trying to achieve as they perform their design calculations, and they know how to measure the results as they evaluate their computer simulations and test their prototypes.

The 777 on which I flew was no exception. Its design embodied a set of specifications negotiated between Boeing and the airlines that would contract to buy the aircraft. The airlines asked Boeing to draw on the technological advances made in the twenty years or more since the wide-bodied Douglas DC-10s and Lockheed L-1011s in their fleets had been designed. Their plans called for a new wide-bodied aircraft with capacity to carry three to four hundred passengers on flights of five thousand miles and more, nonstop flights capable of crossing the Atlantic or Pacific. The airlines wanted to deliver passengers speedily to their destinations in comfort and convenience, and economically, allowing ticket prices to be set lower than on existing aircraft. They insisted on an aircraft so reliable that the Federal Aviation Administration would certify it for transoceanic flights immediately upon entry into commercial service. Other specifications also applied, for the aircraft needed to fit into the airlines' existing infrastructure. The 777 had to take off and land on existing runways and had to load and unload at existing terminal gates. The airlines wanted to continue using existing equipment for refueling, baggage handling, and other tasks. In short, they were not willing to scrap their investments in auxiliary equipment to accommodate the new plane.

Meeting these specifications required more than five years of intense engineering effort, beginning well before Boeing committed to build the 777 in 1990, and culminating in May 1995, when the first paying passengers boarded the aircraft. The engineers proceeded from their conceptual understanding of a jet airliner's architecture and functional principles through the first stages of design, sometimes called project definition or preliminary design. They set the dimensions of the cabin to carry the required number of passengers and from that worked out the size and shape of the fuselage. Extrapolating from earlier designs, the engineers could then make a first estimate of the aircraft's weight, including passengers and fuel for transoceanic flights. In stable flight, lift must equal weight. Since lift is proportional to wing area, they sized the wings' area to obtain the needed lift. With the size and shape of the aircraft approximately determined, they estimated its resistance to airflow—its drag—at the projected cruising velocity and from that determined the thrust that the engines must provide to overcome the drag.

Design is an iterative process, particularly in its preliminary stages. Thus the engineers used computer models, wind tunnel tests, and data from earlier aircraft to refine their calculations and improve their estimates. They streamlined the aircraft's shape and smoothed its surfaces to reduce drag. They optimized the shape of the wings to increase the ratio of lift to drag. They engaged in ongoing searches for opportunities to reduce the aircraft's weight, since more weight meant that either less fuel or fewer passengers could be carried. And they sought engines with greater fuel efficiency, for higher efficiency resulted in lower fuel costs and reduced the amount of fuel that the aircraft would have to carry.

Specifying the number of jet engines was the salient decision in the 777's design. Boeing engineers found it advantageous to reduce the three or four engines used in earlier wide-body jets to just two, each more powerful than any heretofore built. Two engines translated into lower maintenance expense and greater fuel efficiency, and these economies were essential to meet the airlines' targets. The two-engine design also imposed a stringent reliability goal. For before passengers could fly, the engineers had to demonstrate to the safety authorities that the chance of engine failure was so remote that the two-engine 777 would be safer than the four-engine aircraft already flying transoceanic routes.

The trade-offs inherent in determining the number, thrust, and operating characteristics of the 777's engines are indicative of those that permeate the process of engineering design. The designers must seek to satisfy the criteria that their clients set or that marketing studies indicate their customers want. They also must meet the criteria of safety agencies, environmental authorities, and the like. The resulting specifications are multiple and sometimes at odds with one another. Inevitably, difficulties arise in meeting all of them. Long-term technological progress may mitigate these problems, as research leads to engines, for example, that are cheaper, more powerful, less polluting, quieter, and more reliable. But in the short run many such decisions have no answer in further research, for engineers must complete their tasks with the technology at hand and within the time available. As design progresses, the designers must go through the painful process of trade-off and compromise to create technology that comes closest to resolving the conflicts and meeting the design objectives.

An aircraft's designers must move from the most general specifications of size, lift, weight, thrust, and drag to increasing levels of detail. They must place and size the control surfaces in the tail and wings to allow the plane to climb, turn, and descend, and to remain stable in the presence of wind gusts, an engine failure, or other upsets. They must reconcile design requirements for its low-speed behavior as it lifts off from the limited lengths of existing runways with the requirements for efficient high-altitude flight at nearly the speed of sound. The engineers must size and locate the fuel tanks and landing gear, they must place cargo and passenger doors, engine mounts, and more.

Design concepts, the specifications, and the trade-offs work their way down to those who design at ever-more-detailed levels. The 777's engineers embodied the trade-offs between passenger capacity and flight length and those between thrust, drag, lift, and weight in millions of components and parts. Boeing purchased many of them: standard cables, valves, pumps, piping, switches, and the like, and also the electrical devices, mechanical mechanisms, and electronic hardware manufactured by subcontractors specifically for the aviation industry. The engineers also designed hundreds of thousands of parts with unique sizes, shapes, and functions for the 777 and had them fabricated from metals, composites, polymers, and glass. Hundreds of engineering teams, working for Boeing as well as for subcontractors, reduced the concepts and requirements down to the specific dimensions, shapes, and material compositions for the individual parts. They detailed the parts for cargo door latches, rudder controls, engine cowlings, landing gear actuators, emergency light fixtures, and the list goes on and on.

The 777 engineers occupied the forefront of the advances brought by the application of computer technology to the design process. They used a computer-aided design network of thousands of computer workstations and did away with the many thousands of pages of paper drawings previously needed to specify an aircraft's design. Their "paperless design" existed in the memory of the computer network, allowing them to make improvements, specify changes, and correct problems without redoing entire drawings. Other delays were averted, since changes became instantly available on the computers of design teams working on related systems. The computing system aided analysis as well as design, as computer-aided-engineering software checked stresses, temperatures, currents, and other variables to assure that they were within the specified safety margins.

Even the best of computing facilities and scientific resources, however, do not eliminate the need for tacit knowledge. As designers fixed the shape, size, and composition of each individual part, they traded off weight, cost, strength, and other properties. Like the wheelwright of old, whose intuition guided the shaving of unnecessary wood from low-stress regions of hub, spokes, and rim to reduce a wheel's weight without weakening it, the aircraft engineers reduced weight and cost by eliminating unnecessary material from each of the many parts they designed. But unlike the wheelwright, they had computer simulations to determine stress distributions and colorful graphic presentations of the results to indicate where metal might safely be removed and where it must stay. Across the many design teams, engineers applied other tacit knowledge and performed other calculations to assure that machine parts, electrical components, and structural members wouldn't break, leak, burn out, or otherwise malfunction even during the most extreme conditions encountered in flight.

As in all design, creating an airliner is a communal task. Engineers must collaborate closely with industrial designers to assure that ticket-buying passengers will find the aircraft's interior to be pleasant, convenient, and comfortable. They must interact with production and maintenance specialists to ensure that in optimizing each component or part they haven't made it more difficult—and expensive—to manufacture, install, or maintain. They must consider "what if" scenarios to evaluate the dangers of corrosion, cracking, and other modes of failure. Engineers must confer with people from multiple design teams and from numerous professional disciplines, for in an airplane—as in a ship, automobile, or other complex technological system—many different subsystems interact and must fit together in confined spaces.

Within the 777, exceedingly little space remains unused. Hidden from the passengers' view—beneath the floor, above the ceiling, and in the walls—engineers fitted a maze of components and systems: electrical cabling and motors, hydraulic lines and actuators, air ducts, pumps, valves, latches, and more. The computer-aided design system was invaluable in sounding the alarm when two design groups overlapped different objects in the same space. But sounding alarms doesn't solve problems; those responsible for the interfering subsystems conferred and debated. They decided which system would go where, how much space it could take, which must have more ventilation, and which must be more accessible for maintenance. All such conflicts required resolution before the design and eventual construction could go forward.

❀   ❀   ❀

Deepening scientific understating, combined with the power of the computer, has brought forth seminal advances in the knowledge, tools, and procedures at engineers' command. Increasingly, they can isolate, analyze, and eliminate design problems before construction begins. Computer experiments allow the analysis of many alternative designs to optimize performance, decreasing the time and costs required to arrive at acceptable solutions. These computational methods lessen dependence on scale models, mock-ups, and testing and reduce the number of prototypes needed to evaluate alternatives.

Engineering hubris, however, is held in check by the knowledge that, to a greater or lesser extent, each new design project is an adventure into the unknown. Likely, there are holes in the theory, uncertainties in material properties, and other drawbacks to placing too much faith in what can be calculated. As the complexity of technological systems grow, parts and components interact increasingly in unexpected ways and behave in unanticipated manners. Even with application of the most modern of analytic methods, a degree of uncertainty remains. Unwanted surprises are apt to appear. Will each component perform as it should, regardless of how long it is used or how much it's abused?

Engineers verify theory, conjecture, and calculation with physical reality to the greatest extent possible before they put their designs at risk. Thus, long before Boeing constructed the first operational prototype—before the first 777 lifted off the runway—its engineers built numerous models, ran extensive experiments and tests, and gathered and analyzed troves of data on components, subsystems, and materials. They examined scale models of the candidate's fuselage, wing, and tail designs with thousands of hours of wind tunnel experiments to validate and calibrate the computer simulations upon which they relied. They tested materials, components, and subsystems to eliminate uncertainties in performance and durability. They ran motors and exercised valves under extremes of heat and cold. They covered prototype passenger doors with ice to ensure that they functioned properly under extreme conditions and simulated flight conditions to verify the behavior of hydraulic subsystems, electronic controls, and more.

The more complex the component was, the more extensive the test program required. Testing of the 777's engines, which three different manufacturers supplied, began before any other parts. Engines from each manufacturer underwent testing, on the ground and then in flight. Engineers attached a Pratt and Whitney engine to an older 747 test-bed aircraft to observe its performance through the range of flying conditions that would mimic later service. The engines were subjected to abuse to determine their limits—run after being set in an unbalanced state to induce vibrations as well as under extreme conditions of heat and cold. Technicians injected the corpses of dead Canada geese into the engines' intakes at high speed to demonstrate that even if the 777 flew through a flock of large birds, its engines would still produce sufficient thrust for safe flight. Perhaps most stringent of all were the durability tests that the Federal Aviation Administration (FAA) required the engine manufacturers to perform as reliability demonstrations.

The agency's policy reflected the widespread view that the chance of a twin-engine aircraft crashing before it could make an emergency landing was greater than for an aircraft with three or four engines. It required that until a year or more of commercial service had proven engine reliability, a twin-engine airliner could venture no farther than one hour's flying time from the nearest suitable landing site. Indeed, Boeing's earlier twin-engine aircraft, the 767, was in commercial service for two years before the FAA allowed it to travel the three hours from the nearest airport necessary to fly transoceanic routes. Experience, however, indicated that since the regulation was promulgated decades earlier, aircraft engine reliability had increased dramatically, to the point that twin-engine airliners were now more reliable than the three- and four-engine aircraft of earlier generations. Thus Boeing based the 777's design on the premise that immediately following delivery, the airlines could fly it on transoceanic routes. To meet this expectation, the engine manufacturers and Boeing had to demonstrate the improved engine reliability to the FAA before delivering the 777 to the airlines.

The firms undertook this task by putting the engines though comprehensive endurance tests, first on the ground and then on prototype 777s. The ground tests required three thousand trouble-free cycles, each cycling the engine's components through the temperature, pressure, and stress cycles that they would encounter in takeoff, flight, and landing. One engine failure did occur after a few hundred cycles, requiring engineers to redesign the offending components before they restarted the test for an additional three thousand cycles. Before the delivery to the airlines, 777s equipped with each of the three manufacturers' engines underwent one thousand flight tests, simulating two years or more of commercial service, to further validate engine reliability.

The initiation of tests on an operational prototype is a milestone in any design process, for it represents the first chance that engineers have to observe how their creation behaves as a whole. They are called verification prototypes because they are meant to verify the performance, reliability, and durability predicted by the engineers' design. They also reveal weaknesses and idiosyncrasies not apparent in computer simulations or component testing, models, or mock-ups and allow the designers to make adjustments, fine-tune the design, and correct flaws and shortcomings before the product is put to use.

The size and complexity of a large commercial airliner as well as the cost and effort that go into its design make the flight of the first prototype an extremely important event. The corporation contracted to construct it has ventured its future on the success of the aircraft, and if serious weaknesses remain hidden and unresolved at this point, that future could indeed be bleak. The testing that lies ahead must convince safety regulatory bodies, airlines, and the flying public that the aircraft is both safe and economical. In all, Boeing used nine 777s in prototype testing. But following test completion the majority of them were delivered to airlines for commercial service.

Uncertainties loom as a new airplane lifts off the runway for the first time, and emergencies must be anticipated. On the 777's first flights, crewmembers wore parachutes and had planned escape routes, should the worst occur. They first verified the aircraft's performance under normal conditions of takeoff, cruising, and landing. As data were gathered, problems fixed, and confidence gained, the test pilots put the aircraft through more severe tests, pushing the envelope beyond situations that would occur rarely even in a lifetime of usage. They flew the prototypes at altitudes higher and lower than those allowed in passenger service; they deliberately put the aircraft into steep dives and other maneuvers to demonstrate its ability to recover. They flew the aircraft with one engine shut down, and they took off and landed in frigid weather, in desert heat, on flooded runways, and more.

Two of the 777s, however, never flew. Over a period of a year, technicians twisted and stretched the fuselage, wings, and tail of one prototype beyond the worst that bad piloting or wind battering could expose it to. They subjected it to stresses 50 percent greater than the worst that could be expected in a flight emergency, as the FAA required, and then determined how far they could bend the wings before they snapped. They performed fatigue tests on the second prototype, flexing its airframe through the number of stress cycles that the aircraft would experience over its entire lifetime with a 50 percent safety margin, thus sixty thousand instead of forty thousand stress cycles.

The 777's flight testing encompassed nearly five thousand takeoffs and landings as well as more than six thousand hours of flight. Test personnel uncovered difficulties in the course of prototype testing, mostly small but some that were more substantial. The 777 flights exposed problems: vibrations developed in the doors of the compartment housing the nose landing gear, for example, and an inadequate valve design was revealed to be the cause of a cabin depressurization. Design modifications eliminated these faults and others as well, and the completed test program verified that the engineers had met their design goals. The FAA certified the 777 for transoceanic flights, and passenger service began.

⊛　⊛　⊛

The basics of the process followed in engineering the 777 are near universal traits of design: the melding of concepts with requirements at a whole-system level, the progressive refinement at more and more detailed levels of design, and the exhaustive testing and checking to remove uncertainties and validate assumptions. But within this framework, the methods, procedures, and culture of design organizations vary as much as the goals pursued by their engineers. The emphasis on what is important differs from one product to another and from one sector of the economy to another. It differs among aircraft and mass-produced automobiles, appliances, and other consumer products. Likewise, it differs among the engineering of one-of-a-kind bridges, buildings, and dams, as well as in complex integrated systems ranging from air traffic control to waste and water treatment. The differences are apparent even within aeronautics: we can see them by comparing the engineering of a commercial airliner with that of a military aircraft.

Tensions always exist between multiple performance criteria, costs, and dependability. How designers weigh priorities to make acceptable compromises depends strongly on the nature of the product and the use that it serves. The engineers who design commercial airliners face much different priorities from those designing combat aircraft for the military. In military hardware the emphasis must be on performance, whereas cost, and often dependability, must be made subservient. A combat aircraft must be able to fly faster and higher than that of a potential adversary; it must also be more maneuverable and carry the most advanced weapons as well as electronics for detection, evasion, and fire control. If the aircraft's performance does not outclass the enemy's, the number of losses in combat may negate any benefit to its mission, with a concomitant loss of life. Economy in construction and dependability in operation count for little without superior performance.

Maximizing performance in military aircraft necessitates lighter weight, higher stress levels, smaller safety margins, and more complexity than found in civilian aircraft. Operating close to the edge of proven technology inevitably increases the number of break-downs—and even of crashes—that must be tolerated by the military. In contrast, the engineering criteria for civilian aircraft place less emphasis on superior performance relative to cost and comfort, for ticket price and profits are central considerations. And commercial aviation places much more emphasis on safety, for the risks are incurred not by those who become combat pilots, sit in ejection seats, and wear parachutes but by the hundreds of ordinary civilians who occupy each airliner. No aircraft manufacturer can survive once the flying public loses confidence in the safety of its products. Safety is the overriding design consideration; degraded speed, payload, fuel economy, and other performance measures are accepted in order to reduce the chance of an airliner crash to an absolute minimum.

Historically, the quest for superior performance has caused military designers to be first in reaching for new, untried technologies. It is no accident that the jet engine, swept-back wings, and numerous other innovations first appeared on military aircraft. The military could better tolerate the higher costs and poorer reliability as their engineers progressed through a learning curve of applying new technology. Conversely, the overriding concern with safety and economy causes commercial aircraft designers and their airline customers to stay much closer to the tried and true. This conservatism is observed in the similarity in Boeing airliners designed over a twenty-year period, as shown above in figure 47; these in turn appear not that much different from the 707 and 727, designed in earlier decades. In contrast, figure 48 shows the diverse conceptual airframe designs proposed for the fighter plane that was to become the F-22. These designs from the early 1980s, while the cold war still raged, were for an air dominance fighter, and the designers contemplated radical departures from earlier aircraft as they reached for superior maneuverability, longer range, and higher supersonic speeds.

The 777's fly-by-wire system, which transmits the pilot's commands by wire to the control surfaces on the aircraft's wings and tail, exemplifies a technological innovation adopted by the military long before it was judged safe and cheap enough for commercial aviation. Before fly-by-wire, cables guided by pulleys mechanically

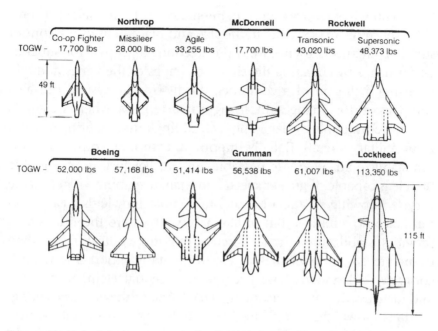

	Northrop			McDonnell		Rockwell	
	Co-op Fighter	Missileer	Agile			Transonic	Supersonic
TOGW –	17,700 lbs	28,000 lbs	33,255 lbs	17,700 lbs		43,020 lbs	48,373 lbs

49 ft

	Boeing		Grumman			Lockheed
TOGW –	52,000 lbs	57,168 lbs	51,414 lbs	56,538 lbs	61,007 lbs	113,350 lbs

115 ft

Figure 48. Varied design concepts for fighter aircraft. (From David C. Aronstein, Michael J. Hirschberg, and Albert C. Piccirillo, *Advanced Tactical Fighter to F-22 Raptor: Origins of the 21st Century Air Dominance Fighter* [Reston, VA: American Institute of Aeronautics and Astronautics, 1998], p. 41. Copyright 1998 by the American Institute of Aeronautics and Astronautics, Inc. Reprinted with permission.)

linked the cockpit to an aircraft's control surfaces. The movement of these control surfaces channeled the airflow over the wings and tail such that the pilot controlled an aircraft's climbs, turns, and other maneuvers through the forces transmitted from the cockpit control stick and pedals to move the ailerons, stabilizers, and rudder. Such systems are reliable and durable, but the aircraft's design must be sufficiently stable to allow the pilot's reflexes to easily compensate for wind gusts and other disturbances.

Fly-by-wire systems transmit commands from cockpit to control surfaces by electrical signals instead of mechanical cables. Such a system has a number of advantages. Replacing cable-and-pulley systems with wires reduces both maintenance cost and weight. It provides much more flexibility in the size and placement of the control surfaces. Engineers then have more freedom to design for superior performance by changing the size and shapes of the wings and tail. With the inclusion of a computer, the fly-by-wire system can also compensate for changes in flying characteristics when weight is shifted, for example by depleting fuel, adding cargo, or firing missiles from military aircraft. Equally important, computer control systems can stabilize an otherwise unstable aircraft. For just as an unstable bicycle is capable of greater speed and maneuverability than a stable tricycle, allowing an aircraft design to be unstable yields greater performance possibilities. But unlike a bicycle, where the rider's skill provides stabilization, a jet aircraft requires control faster than human response times, making electronic stabilization crucial. Such computerized systems have become an essential element in flying and maneuvering at supersonic speeds. None of the airframe configurations shown above in figure 47 would be viable using mechanical cable control systems. These aircraft sacrifice stability for performance, and in doing that, they must rely on highly sophisticated fly-by-wire systems, at the center of which are powerful digital computers.

The air force initiated fly-by-wire in the 1960s, and soon afterward NASA initiated its own program. Both programs experimented by modifying existing fighter planes, an F-2 and F-4 respectively, and examined two classes of systems—analog and digital. At first the air force concentrated on analog systems, in which the controls are embedded in circuitry, without the need for a computer or its software programming. NASA focused on the increased performance potential of digital systems, with the first flight of its modified F-4 taking off in 1972. Research overcame many challenges, but like many technological innovations, fly-by-wire was costly at first, compared with what it replaced. But with additional experience costs came down and reliability improved.

In their quest for performance, the armed forces were first to adopt fly-by-wire. Next, engineers employed it in the supersonic Concorde. But more than a decade elapsed before its adoption became more widespread for commercial aircraft. When Boeing introduced the 757 and 767 in the 1980s, its engineers relied on the tried-and-true mechanical controls, even though by then digital fly-by-wire controls had become the nearly universal choice for military aircraft. The delays resulted from the need for the FAA, its international counterparts, the aircraft manufacturers, and the airlines to be convinced that the reliability of fly-by-wire was sufficient for use in commercial airliners. Lingering doubts related to computer hardware failures, software glitches, and other potential crash causes had to be resolved before commercial application of fly-by-wire could proceed. The military and NASA had indeed worked out systems of redundant and backup computers, duplicate sensors, fault detection, and fault-tolerant systems so that even if equipment failure or software faults occurred, the pilot would not lose control of the aircraft. But still, Boeing gave fly-by-wire consideration for commercial airliner design only after years of military operation proved the system's robust capabilities.

Fly-by-wire finally achieved success in commercial aviation, first in the Airbus 320, introduced in 1988, and then in the Boeing 777. Capitalizing on military research, development, and experience, the 777's system emphasized both redundancy and diversity with not one computer but nine! The designers arranged them in voting configurations to compensate for possible equipment failures. Moreover, each of three groups of computers used chips manufactured by different corporations and software programmed in different languages to preclude the chance that an equipment failure, programming bug, or procedural error could disable the system.

The extent of such design and testing precautions and the need to satisfy governmental safety agencies indicate why the design culture of commercial aviation was much slower in adopting fly-by-wire technology than that of the military. Similar histories apply to other innovations, not only for aircraft but across the technological spectrum. The balancing of design concerns differs between civilian and military ventures as well as between cultures, industries, and corporations. For as engineers weigh the importance of performance, economy, and dependability, the balances they strike must be as varied as the clients and customers they seek to serve.

Ten

Heading for the Road, Reaching for the Sky

The view is truly panoramic from atop Chicago's hundred-story John Hancock Center. To the east, the expanse of Lake Michigan stretches to the horizon, the sky broken only by an occasional silver speck descending toward the airfield at O'Hare. The other points of the compass display the breadth of industrialized society: to the west and south stand the Sears Tower and the city's other architectural landmarks; to the north and far below, two ribbons of automobiles wend their way up and down Lake Shore Drive. The view of the skyscrapers and automobiles brings to mind a panorama of another kind, a panorama of engineering that ranges from the design of autos and other goods mass-produced by the million to projects culminating in monumental one-of-a-kind creations such as the Hancock Center. The creation of automobiles and of skyscrapers differs in important respects, and it differs again from the aeronautical achievements at O'Hare Airport. But beneath the wide range of technology that engineering creates, a core of underlying themes pervades the profession.

Engineering is at the heart of both the production of automobiles and the construction of skyscrapers, but the challenges encountered differ significantly. Those who engineer mass-produced products rely on custom-built prototypes—dozens or more—to gain user input and to test their creations under diverse and extreme conditions. But then they must replicate the prototype's performance in the mass-produced item at a cost low enough that will make it feasible for the public to buy. In contrast, the builders of skyscrapers face challenges more closely allied to those found in the engineering of bridges, dams, refineries, or networks for supplying communications, water, or power. If a project's product is one-of-a-kind, prototypes cannot be built, and designers must substitute other means to gain the confidence and knowledge necessary for success. Moreover, intermediate between the engineering of mass-produced and one-of-a-kind products are aircraft and similar technologies. They are produced by the dozens or hundreds, with engineering methods to match. The contrast between producing automobiles and building skyscrapers highlights these differences and exemplifies the diversity of the engineering enterprise.

The conceptual design that defines an automobile, like that of an airplane, has long been established, dating back to Henry Ford and before. But unlike an airplane, automobile specifications do not emerge from complex technical negotiations between the builder and airline executives or high-ranking military officials. They result from corporate efforts to determine what ordinary citizens value most in their purchases and how much they are willing to pay. Like other technologically mature products sold on consumer markets, automobiles' technical performance capabilities—speed, acceleration, distance between refueling, and design life—tend to exceed the needs of most buyers. Purchase price and safety, of course, remain central considerations. Even more, the visual image that the auto creates, the style and comfort its interior, the feel of its drive, and the convenience afforded by its features become dominant determinants of success.

Visual appeal, comfort, and convenience are more elusive than the numerical criteria with which the engineers' analytic minds and quantitative methods are most comfortable. Thus the design of automobiles—and that of many other consumer products—flows first from the artistic conceptions of industrial designers and from the studies of psychologists, ethnographers, and others on how product and buyer are likely to interact.[1] Artistic intent, coupled with aerodynamics, takes center stage in determining the lines, contours, and profile of an automobile. The designers sketch and draw, and then they sculpt full-size models in clay, such as the one shown in figure 49. Marketing experts, car dealers, consumer focus groups, and auto executives pass judgment on the auto's consumer appeal, and their opinions further influence the details on the car's lines, shape, and interior.

Figure 49. Sculpting a full-size clay model. (Reproduced by permission of the General Motors Archive.)

But artistic intent and engineering reality must be melded into a unified whole. If design is to go beyond strictly trial-and-error efforts, engineers must collaborate closely with industrial designers as well as sales and marketing experts to reduce human preferences to objective criteria. Reducing the feel of the steering, the grip of the brakes, the stiffness of the suspension, and the comfort of the seats to numerical targets may be difficult, but it greatly enhances the engineers' ability to determine what they are designing and to measure their degree of success. Between an auto's sleek exterior and its comfortable interior—in the confined spaces out of the buyers' sight—the engineers must squeeze the engine, suspension, steering, braking systems, and much more, all designed to criteria that will meet with the public's approval.

As in most technological endeavors, an automobile's design is a communal effort. It entails argument, conflict, compromise, and resolution as well as drawing, computation, testing, and experiment. Disagreements break out among electrical, mechanical, and industrial engineers as well as between those responsible for styling, steering, braking, and power transmission. From the limited air flow into a cramped engine compartment, for example, engineers must agree on the air allocated to engine cooling and passenger compartment ventilation, bringing those responsible for engine performance and for passenger comfort into conflict. They must work out compromises to accommodate electrical cabling, cooling ducts, and fuel and hydraulic lines when they interfere with one another.

Production engineers object to body styles that require bending sheet metal into shapes that are impossible to obtain at reasonable cost without creating ripples, cracks, or other defects. Manufacturing engineers take issue with mechanisms that will be difficult—and therefore expensive—to assemble, as do maintenance specialists with the placement of parts that make them difficult to service or replace. Such challenges are compounded by requirements to meet other criteria, criteria that are sometimes in conflict. For example, meeting fuel efficiency standards by shaving an auto's weight is likely to cause added difficulty in satisfying crashworthiness standards.

Often these trade-offs are safety related. For whether an automobile is being made more crash-resistant or a building more earth-quake-resistant, added strength can require added material, and added material comes at a price. In automobiles, as in airplanes, ships, or practically anything that moves, the added weight that accompanies added strength reduces fuel economy and decreases distances between refueling points. Thus engineers grapple with the problems of minimizing weight and costs, optimizing performance, and providing a design that meets safety requirements.

At some point requirements for further risk reduction will render a product too expensive or degrade its performance so much that it is no longer feasible. Before that point is reached, decisions must be made as to whether the technology's benefits to society outweigh its risks. Abstract issues of the risks and benefits of technology must be reduced to concrete design rules and regulations that allow engineers to make the myriad of decisions required to design and produce the technological entities that support our standard of living.

Decisions to increase strength and reliability, to add safety features, and to provide redundant or backup systems pit performance and cost against risk reduction. For the most part, such design refinements lead to net safety enhancements. However, some decisions made with the intention of improving safety may reduce the likelihood of one risk but increase that of another. Risks may affect different people in different ways, and situations arise in which design decisions reduce overall risk but at the same time shift risks from one user to another. Consider, for example, the dilemmas faced by those responsible for the design of the safety air bags on our automobiles.

Air bags represent a significant advance in automobile safety. But even there, the design engineers encounter trade-offs and value judgments that tightly tie highly technical considerations with matters that are essentially social and political. Since their initiation, air bags have saved thousands of lives and reduced the severity of injuries in serious collisions. But they are not without problems. These stem from two important parameters that the designers must set: the level of impact required to trigger air bag inflation and the speed at which the bag inflates.

Raising or lowering the impact level that will trigger the bags results in mixed consequences. A lower triggering level is surer to offer protection from all accidents. But in less severe accidents, the air bag's inflation force may cause injury worse than those incurred from the accident. Conversely, a higher triggering threshold may allow too many injuries from impacts that are not strong enough to trigger the bag. Unfortunately, no triggering level exists at which the former problem is lessened without aggravating the latter. Designers pick a level that attempts to minimize the total number of injuries. But their decision will not prevent lawsuits from those injured in accidents too mild to trigger air bag inflation and those harmed by the deployment of the air bag itself.

The force of high-speed bag inflation has killed numerous children and small adults. Designers had set that speed high enough to prevent injury to an adult who was not wearing a seat belt. If the engineers were allowed to assume that adults would wear seat belts, then they could have reduced the inflation speed, also reducing the chance of injury or death to smaller passengers. Thus the designer faces a social question: should the inflation speed be chosen to optimize protection for the many adults and teenagers who refuse to buckle up or to provide maximum protection of the smaller number of child passengers?

Eventually, the dilemma may be resolved. Future air bag systems may be more complex, coupling sensors to determine the passengers' weights and whether they have seat belts fastened with electronic controls to adjust the air bag inflation speed accordingly. But with these more complex systems, the frequency of bag failures may increase, for sensor and adjustment mechanism malfunctions must also be considered. Added cost will also be a consideration, for higher prices may dissuade increasing numbers of motorists from selecting air bag options. But if the safety improvements are large enough, government regulatory bodies may impose that price by mandating air bags. Once again detailed design decisions are in fact not only technical but also political and social in nature.

Air bags, of course, are not unique in raising complicating trade-offs in determining adequate levels of safety. Questions arise in weighing the risks from one class of failure with another: specifying a lower tire pressure increases traction on slippery pavement, but it also increases wear and the likelihood of catastrophic failure in high-speed long-distance travel, particularly in hot weather. The effectiveness of air bags, better tires, and other safety improvements must also be weighed against the offsetting propensity of drivers to behave less cautiously as they gain confidence in the safety of their vehicles.

These trade-offs, like a myriad of other decisions made in design, mix technical, social, economic, and political considerations in complex combinations. How society deals with technological risk and determining how safe is safe enough involves design organizations, governmental regulatory bodies, insurance underwriters, and professional standards and codes. The processes are far from simple, and they often seem at odds with the simple ideal of minimizing the total number of deaths that accompany the benefits of technology.

As a design progresses, compromises are made and issues resolved. Meanwhile, engineers must also concentrate their efforts on predicting how their design will behave. They perform computer simulations, build models and mock-ups, run engines, and test components. They subject their models to wind tunnel testing, in which the forced airflow around the model allows them to corroborate their predictions of the auto's air resistance, wind noise, and other variables. They attempt to provide real-world environments by testing new brake, steering, or other mechanisms by installing them for testing in existing automobiles. They use whatever means they can devise to reduce to a minimum the chances of a major design misstep. For next they will undertake the expense of building operational prototypes to road-test their newly designed automobile.

The contrast in prototype construction between aircraft and automobiles is marked. For whereas Boeing used only nine 777s for prototype testing, most of which were subsequently delivered to airlines for passenger service, automakers typically build three to four hundred prototypes for testing a design, and few, if any, survive to be sold. One hundred or more will be smashed—front, side, and rear—to verify the auto's ability to meet crashworthiness standards. Others undergo durability tests, being driven fifty thousand miles or more through mud and salt water, over gravel, potholes, cobblestone, broken glass, and other test track obstacles to simulate wear and tear well beyond what most cars that are sold should experience in a lifetime that may exceed one hundred thousand miles. Some prototypes undergo testing in extreme environments, in summer desert heat and in arctic winter cold, to stress heating and air conditioning beyond normal use. They are sprayed with salt water, covered with ice, driven through floods, and more.

Engineers find prototypes invaluable for identifying subtle problems that are difficult to predict with component testing or computer simulations, such as unexpected sources of road noise, squeaks, and rattles. But they must track down these problems and then make design fixes to eliminate them. Even more of a worry are design flaws with safety implications, which if undetected and uncorrected could result later in massive recalls, after many thousands of the autos have been sold. Major flaws and fundamental problems do turn up; some may be corrected with simple fixes, but others may be more serious. Redesign may then require rebuilding major parts of the prototypes and then retesting them to ensure that the problem has been eliminated. Such setbacks stretch development budgets and overrun schedules that must be met if the auto is to go into production and reach the buyers by its target date.

The prototypes for automobiles and other mass-produced products differ from those of the 777 not only in number but also in how they are built. Teams of 777 engineers designed the manufacturing tools, equipment, and system simultaneously with the aircraft and used them to produce both the nine 777 prototypes as well as the aircraft to follow. As prototype building and testing progressed, they corrected weaknesses identified in the details of both production methods and the aircraft's design. At completion of prototype testing, engineers finalized aircraft design and refined their manufacturing methods as they prepared for production to commence.

Automobile prototyping follows a pattern more common to consumer goods produced in truly large numbers. As an auto's design progresses, production engineers plan for the machine tools and robots as well as the organization needed for high-speed assembly line throughput, turning out cars at one per minute or more. The detailed organization, tooling, and staffing of such lines cannot be completed, however, until engineers finalize the auto's design. Conversely, design engineers must have the prototype autos available for testing well before mass production facilities are ready to operate. Thus engineers build a small pilot plant where highly skilled technicians and trade workers craft prototypes one at a time, without the aid of robots or assembly line organization. Some prototype parts are off-the-shelf, mass-produced items: the standard nuts, bolts, batteries, spark plugs, and so on. But many are unique to the design, produced by job shops in small numbers specifically for the prototypes. The pilot rate of production is closer to one per day than the one-per-minute target of an assembly line, which translates into a large cost per vehicle. For the mass-produced family sedan selling for $20,000 or less, the cost of each handcrafted prototype may be $250,000 or more.

Progressing from pilot plant prototyping to assembly line production, engineers must ensure that defective automobiles don't result from poorly adjusted machinery, inattentive line workers, or a supplier's defective parts. Just as design flaws are the design engineers' nightmare, manufacturing defects are the production engineers' nemesis. The engineers must ensure that high-speed parts fabrication and assembly operations, carried out by robots and hourly-wage workers, faithfully replicate the automobile that emerged from the careful crafting of the prototype pilot plant. They face an added challenge: thorough test drives to verify that each automobile is produced as it was designed would be far too costly. Produced in small numbers, aircraft may undergo individual certification flights before passengers step aboard. In contrast, high rates of production allow sampling of only a small percentage of the finished automobiles to verify that their performance is meeting the standards to which they were designed. Thus defects must be detected and eliminated early, before the autos roll off the assembly line. Such are the challenges of producing technological products in large numbers at prices that the public can afford.

One-of-a-kind projects face several challenges that are quite different from those encountered in engineering automobiles or other mass-produced products. Dams, bridges, buildings, and other massive structures are erected not in the controlled environment of a factory, but outdoors, subjected to all of the weather's vicissitudes. Thus construction quality must undergo constant and rigorous checking to confirm the consistency of the concrete, the anchoring to foundations, and the joining of prefabricated beams and columns. Geological surprises encountered as foundations are dug, labor or materials shortages encountered during construction, or shifting client requirements may call for modifications in design details or construction procedures between ground breaking and project completion. But all involved must take care that on-the-spot changes to improve design or facilitate construction are not adopted without their implications being thoroughly analyzed. Collapses and other ruinous failures have resulted from such ad hoc modifications, made in the rush of construction schedules without checking to ensure that design integrity is maintained.[2]

The need for long design lives of many one-of-a-kind structures in combination with the environments that will confront them pose formidable challenges. Whereas an automobile may be expected to last ten or fifteen years, or an airplane twenty or thirty, the design life of large structures may be a hundred years or more. Thus they must be designed to withstand extremely rare events, including earthquakes, floods, tornadoes, and other natural phenomena of a severity that may occur centuries apart. The gathering of data and predicting the forces that such rare events may exert is in itself a difficult undertaking.

Lastly, and perhaps most important, one-of-a-kind structures and systems must be designed without the aid of full-sized functional prototypes. Architects and engineers must substitute other means to test and validate their designs before construction begins.

The John Hancock Center—or "Big John," as many Chicagoans know it—offers some insight into the unique challenges faced in a building project of monumental proportions. The Hancock Center originated when real estate developers acquired a prime piece of land along Chicago's fashionable North Michigan Avenue, located a short distance west of the Lake Michigan shoreline. The developers retained the architectural firm of Skidmore, Owings, and Merrill to design a building complex for high-density use of the valuable real estate, which would accommodate retail space, parking, commercial offices, and a sizable number of apartment units. Unlike automotive or aeronautical engineering, where the conceptual design has already been forged through long experience, the designers of the Hancock Center had a great deal of latitude in determining the form that the building or buildings would take. The project began with the developers' specification for eight hundred thousand square feet of commercial space, one million square feet for offices, one million square feet for residential space, and an attractive complex that would draw corporations, businesses, and residents at rents high enough to make the project a financial success.

Any technological project must proceed from a conceptual design; the architects considered two for the John Hancock Center. The first envisioned a pair of buildings, each with a million square feet of floor space. Office space would occupy one, and residential apartments the other. The two towers were to be connected by an eight-hundred-thousand-square-foot building to be occupied by commercial establishments and parking facilities. This arrangement, however, was problematic. As the diagram on the left in figure 50 indicates, the commercial space and parking structure would rise nearly ten stories above street level and extend nearly to the property lines. This configuration would create pedestrian congestion as well as a canyonlike environment that many urbanites find claustrophobic. Moreover, the two towers would be uncomfortably close together, with office windows directly across from residents' apartments.

In search of a more satisfactory arrangement, Fazlur Khan, the project's chief structural engineer, in collaboration with its architect, Bruce Graham, proposed an innovative concept. Khan's idea met the mixed-use space requirements for the property, while at the same time allowing for an open plaza at street level, which greatly enhanced the urban environment. He proposed a single building, the tapered one-hundred-story skyscraper depicted at right in figure

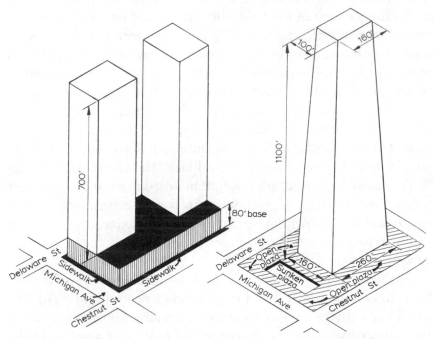

Figure 50. Two conceptual designs for the John Hancock Center in Chicago. (From Fazlur R. Khan, "100 Storey John Hancock Center, Chicago: A Case Study of the Design Process," *Engineering Structures* 5 [1983]: 12. Copyright 1983, reprinted with permission from Elsevier.)

50. The taper, he reasoned, would accommodate commercial, corporate, and residential occupants. Businesses required large floor areas, and if much of the interior was located far from windows, that wasn't a problem. For residential apartments, however, windows are paramount. Thus Khan sought to limit the cross-sectional area of the residential floors of the building, for otherwise the interior space would be exceedingly difficult to rent. Tapering the building resolved the issue. Retail and parking space occupied the first eighteen stories, businesses resided between the eighteenth and forty-fifth floors, and above that there was ample window space for the apartments occupying the structure's narrower upper reaches.

Graham and Khan, architect and engineer, worked hand in glove to bring the Hancock Center into existence. In more routine building projects the architect might be able to complete the essence of a structure's aesthetic and functional design before turning it over to engineers to ensure that it can be adequately supported, and to design the elevator, water supply, power, and other technological systems. The Hancock Center, in contrast, represented a novel structural concept, and thus from the very beginning its design was an intimate mix of engineering and architectural considerations.

Khan and Graham faced many challenges in assuring that the tapered concept would work structurally as well as architecturally. The building's aesthetic impact, as well as its suitability for use, was closely tied to structural challenges. Khan had to ensure not only that the building would support its own weight but that it would be rigid enough to withstand the strong winds at high elevations without excessive bending, cracked walls, or other damage. He also had to anchor the building against settling and make it resistant to the shaking that even rare earthquakes might cause. Moreover, Khan shared with other skyscraper engineers the knowledge that he must design without having prototypes to validate his ideas before construction began. But like others successful in his field, he carried with him a wealth of knowledge of what had gone before, from his own experience, and from the successes and limitations of his predecessors' earlier innovations.

Skyscrapers built well before World War II, such as New York's Chrysler and Empire State buildings, were erected around skeletons of steel beams, columns, and braces. The steel framing supported the weight of floors and also of the walls, which were said to be hung on the steel frame like curtains. These "curtain walls" consisted primarily of stone or brick masonry, with a third or less of their area occupied by windows. As the buildings grew taller, the structural challenges that they presented multiplied. At the skeleton's base, the columns had to be strong enough to support the weight of steel and masonry added with each additional story, and the structure's upper levels had to be stiff—that is, rigid—enough to resist lateral forces of wind gusts that grew more intense as the building's height increased. The mass of the masonry walls provided stiffness against the gusting winds, but for adequate strength engineers had to run massive columns of steel, called core trusses, up through the centers of the buildings, usually adjoining a central core of elevator shafts.

Such designs had their limitations. The massive core trusses robbed the structure of significant amounts of usable floor space, and the masonry walls limited the natural light to the interior. Moreover, architects designed these structures long before engineers could employ computers to perform accurate analysis of their complex forms. The slide rules and hand calculation that they relied on were incapable of estimating stresses accurately. Thus they had to compensate for their uncertainty by using very large safety factors. The buildings were thus overdesigned, resulting in additional masses of masonry and steel, raising the cost of the floor space created.

Advances in materials, the knowledge of structures, and design methods allowed new forms to be employed in the tall buildings of the 1950s. Builders eliminated much of the masonry so prominent in earlier skyscrapers. They greatly expanded window space by hanging vast expanses of glass from steel or reinforced concrete framing. With the weight of the masonry contained in earlier skyscrapers replaced with vast expanses of glass, however, the curtain walls of the newer buildings lent little stiffness to the structures. Without added stiffener to make the steel frames more rigid, the buildings would sway excessively and possibly vibrate on windy days, bringing uneasiness and possibly motion sickness to occupants of the upper floors.

Engineers first met the challenge of bracing these tall buildings of steel and glass against the wind with more massive core columns. This solution was far from ideal, however, for the columns took up valuable floor space as well as increasing the cost of steel necessary for construction. However, a structural innovation in the 1960s, tubular design, alleviated the stiffening problem significantly, allowing architects to design slenderer buildings with smaller footprints, thus allowing them to preserve more land as open space. Tubular design viewed the structure as a hollow tube perforated by windows. Since Galileo's analysis of beams centuries earlier, engineers understood that a thin, hollow tube is much stronger than a solid rod made of the same mass of material. In tubular design, the building's strength came from the rows of columns that formed its periphery, freeing the interior from massive support structures that had robbed earlier buildings of floor space. Larger windows—the curtains of glass—formed vertical rows between the support columns that formed the tubular periphery.

The tapered shape proposed for the Hancock Center, however, presented additional challenges. Unlike conventional tubular design, in which the building's strength was provided by the set of vertical steel columns that formed structure's periphery, the Hancock Center's columns would not be vertical, but slanted inward. That slant replaced more straightforward right-angle connections between beams and columns with connections at smaller angles. More important, the need for strength and rigidity against gusting winds would require that the slanted columns draw so close together toward the top of the building that there would be inadequate space for windows between them. Khan's solution was the innovative cross-bracing, shown in figure 51, that contributes to the Hancock Center's distinctive appearance. His analysis showed that cross-bracing greatly increased the structure's strength and rigidity and distributed the forces of wind and weight more evenly among its columns. The novel design eliminated the need for the space-robbing columns in the core of the building and permitted thinner steel framing and thus more generous window expanses on the building's faces.

Khan's earlier experience served him well in facing the challenges posed by the Hancock Center's unique form. He had been a pioneer in the design of tubular skyscraper structures, applying the principles of this design technique in the construction of several earlier buildings. His design of the nineteen-story steel-and-glass tubular Inland Steel Building was particularly relevant, for it foretold much of what is found in the Hancock Center. Nevertheless, the tapered building of one hundred stories would be a structural extrapolation into the unknown, one that he would face without the aid of prototype testing. Instead, he would rely on computer simulations and the testing of components and scale models. He also performed experiments with human subjects to determine what occupants' reactions would be to buildings swaying from gusting winds.

Figure 51. The John Hancock Center under construction.
Note the X-shaped cross-bracing on the sides of the structure.
(From Mir M. Ali, *Art of the Skyscraper: The Genius of Fazlur Khan*
[New York: Rizzoli, 2001]. Photo courtesy McShane-Fleming Studios;
reproduced by permission of Mir M. Ali.)

To determine how stiff the structure should be, Khan required an estimate of how severe wind-induced motions must become before the occupants of the building's upper floors would become sensitive to the swaying. Since adequate data was not available, he arranged observations and experimentation to gain user input that he then factored into the structure's design. He performed experiments at Chicago's Museum of Science and Industry. He had nine volunteers sit on a large rotating disk, twenty feet in diameter. By accelerating this turntable at various rates and relating the turntable's movements mathematically to those of the building, he was able to estimate how much wind-induced motion the building could sustain before its residents would register discomfort. To obtain corroborating data, he also placed instruments on an upper floor of a nearby twenty-six story steel-and-glass building, recorded its motions during windstorms, and interviewed residents to determine at what point the motions became bothersome.

Although primitive by today's standards, the computers of the 1960s were a great boon to structural analysis. For the first time, they allowed designers to analyze their creations in three-dimensional detail, eliminating most of the simplifications necessitated by the hand calculations of earlier decades. The engineers could then reduce the overdesign that compensated for earlier uncertainty and thus create more efficient structures; they could use computers to analyze forms that previously they would not have dared to attempt. Thus it was Khan who came to the forefront in developing computational methods for analyzing novel structures, applying them to the Hancock Center's tapered structure and to the system of cross-bracing that is inseparable from it. Khan computed the effects of wind and gravity loading on the cross-braced structure and brought it much closer to the ideal, in which each steel member is equally stressed. He pushed the emerging computing methods of the day to their limits, using early IBM equipment and three independent computer codes to cross-check results as he masterminded the creation of the Hancock Center. It would be Chicago's tallest building—but only until Khan designed the Sears Tower, which would be even taller.

Computer predictions and experience with smaller steel-and-glass buildings were both invaluable but insufficient to ensure the success of the structure. Thus as he designed, Khan also had scale models of the Hancock and of surrounding buildings built. Wind tunnel tests on the models then provided understanding of the effects of high wind and wind gusts on the structure and determined how the presence of surrounding buildings would affect the wind forces on the Hancock.

Khan also contracted for extensive load testing: heavy weights—or loads—were placed on structural subassemblies and on the fixtures by which the beams and columns of the building were to be welded and bolted together. The weights simulated the stresses that these building components were expected to experience during the structure's lifetime. Thus the tests ensured that weaknesses in design would be found and corrected before construction was underway. Khan's innovations also allowed more of the fabrication work to be completed in a factory environment of the steel subcontractor, simplifying the on-site assembly of the structure. He thus reduced the number of dangerous tasks that construction workers had to perform from exposed perches hundreds of feet above Michigan Avenue.

The Hancock Center's steel-and-glass construction raised other problems that required attention. Khan analyzed the compatibility of glass and steel under the severe temperature changes that the building's exterior would experience. Inasmuch as metal expands and contracts more than glass, developing a design that prevented glass from cracking or windows from popping out of their frames presented a challenge. Khan's solution was to construct a ducting system that allowed natural circulation to bathe the interior frame and glass year-round in room temperature air.

Of course, much more than architecture and structural engineering go into the creation of a skyscraper. Elevator systems must be designed and optimized, and heating and ventilation systems must closely control temperature and humidity throughout the interior. Complex systems must bring water pressure, electric power, and telecommunications to the top of the building, and waste disposal must be provided. Space must be found for all of the shafts, ducts, pipes, cables, and machinery that these systems require. But they must be shoehorned into the building in such a manner that they are compatible with one another, that they do not detract from the structure's architectural qualities, and that they do not rob it of inordinate amounts of otherwise rentable space.

Even with excellent design, the Hancock Center, like other one-of-a-kind structures, machines, or systems, encountered difficulties along the way. The most serious of these was a problem with the caissons—the massive waterproof structures of concrete that anchor the building's steel columns to bedrock. Chicago is more difficult to build on than New York, for in New York the bedrock comes close to the surface; in Chicago it lies far below. The Hancock Center set a local record, with some of the fifty-seven caissons extending nearly two hundred feet downward. After the structure's frame had already reached a height of more than ten stories, trouble began to appear. Testing and analysis led to the conclusion that one of the building's columns had sunk by one inch. Though seemingly insignificant, that inch indicated a serious caisson problem. Indeed, investigators located a defective caisson, the problem caused by a faulty process for setting the concrete. Six months elapsed while the situation was investigated and analyzed, and the problem was resolved through replacement of the weakened concrete.

With the challenges of design and the problems of construction overcome, Big John rose to bring credit to its creators. Advances in conceptual design and engineering analysis created a building more economical than the skyscrapers of earlier generations. Such economy is sometimes assessed from the weight of steel required per square foot of leasable floor space. By this measure, the Hancock Center required 30 percent less steel than had the Empire State Building. Even more, its aesthetics contributed to the Hancock Center's success: it offered generous plaza space at ground level, creating not canyons but open space. Tailored to the varied needs of commercial, office, and residential occupants, its tapered silhouette became a landmark rising above Chicago's shoreline.

❀   ❀   ❀

Engineers create incredible diversity of devices, products, and systems. They include biomedical, electronic, and nuclear devices, whose conceptual designs stem from recent research findings; they also create airplanes, automobiles, bridges, and buildings, whose beginnings go back for many generations. Their creations range in size from pacemakers and microprocessors to dams and supertankers, and from the simplicity of rivets and screws to the complexity of continent-spanning power and communications networks. They include unique structures of monumental proportions and consumer products produced by the millions. Success often requires that engineers rely heavily on the talents of others, particularly those of a more artistic bent. They work in partnership with architects, who focus on creating buildings that are attractive and attuned to human use. They collaborate closely with industrial designers to ensure that mass-produced products not only function properly and can be built economically but are aesthetically pleasing and easy to use.

No wonder that defining what underlies the engineering of this wide range of artifacts is far from simple. Its diversity notwithstanding, the engineer's profession is unified by the common core of the design process. Customers' and clients' perceived needs must be transformed to specifications to which designs can be specifically targeted. Design must proceed though a hierarchy from general concepts—operating principles and system layout—to the minutest details. Designs must be tested, and the process must cycle back again with revisions, corrections, and optimization until subsystems, components, and parts form a coherent whole. And even though the overall concepts that define the technology may be decades or even centuries old, innovation with new materials, devices, and subsystems are essential to push that conceptual umbrella forward with ingenuity to meet ever-changing needs. And even though similar products or processes may already exist, each project must tailor its design to the particular client's needs or for a particular market niche.

Conflict and compromise are inherent in engineering. They arise in resolving incompatible performance, cost, and dependability criteria. In complex projects they surface in interdisciplinary struggles between stress analysts, circuit designers, heat transfer experts, ergonomic specialists, manufacturing engineers, and others as they press to ensure that the technology for which they are responsible serves its function, that it does not become a system's weak link. Disputes may erupt with architects or industrial designers when technical constraints come into conflict with appearance, aesthetics, or convenience of use. The communal effort goes forward despite budget limitations and deadlines. And overarching the cycles of designing, computing, and testing is the drive to vanquish uncertainty and gain confidence that neither design flaws nor production defects will mar the product of the group's combined efforts.

In carrying forward such efforts, today's engineers rely heavily on the methods and findings of modern science; they utilize sophisticated computer-aided design methods and work in complex organizational structures to meet their goals. But at a deeper level, the engineer's innate prowess is what drives the process. It is their penchant for problem solving, their technological exuberance, that joy in conceiving something new, building it, and making it work. It is their drive to pull things together, devising new solutions and using whatever means available to accomplish their tasks. Such traits often appear early on, in childhood enthusiasm for puzzle solving, model building, and tinkering with everything from computers to chemistry sets. These habits of mind that draw them to their profession also bring the satisfaction and sense of accomplishment that comes as they succeed in creating technology that is strong enough, lasts long enough, and is economical enough to serve society in its intended use.

Eleven

Rocket Science and More

On July 20, 1969, Neil Armstrong stepped from *Apollo 11*'s landing module onto the surface of the moon. His first words, "That's one small step for man, one giant leap for mankind," signaled the climax of one of the great technological feats of the twentieth century. That event, more than any other, popularized the term *rocket science* among the public and engineers as well. It characterized the intellectual hurdles that were overcome to achieve such a momentous accomplishment, the capacity needed to take on intellectually challenging tasks and solve daunting problems. As a corollary, the more widely repeated phrase "this isn't rocket science" came to allude to easily understood problems.

The 1960s space race, which culminated in the *Apollo 11* moon landing, provides a look at how "rocket science" came to encapsulate the growing interplay between twentieth-century engineering and science. The story owes its beginnings to the advances of the 1950s that brought space exploration into the realm of technological feasibility. The large, liquid-fueled rockets developed for military use became capable of lifting payloads into Earth orbit and beyond. And the replacement of vacuum tubes with solid-state electronics brought development of the sophisticated computers, communications, and control systems necessary for space travel to become a reality.

As important to the inception of the Apollo program as technical advances were, however, cold war competition accelerated the drive toward space travel even more. Impatience was palpable in the United States as the Soviet Union forged ahead, launching *Sputnik*, the first man-made Earth-orbiting satellite, in 1957. It added to its triumph by orbiting a dog in *Sputnik II* and hinted that human flight would soon follow. The satellites offered graphic evidence of Soviet technological superiority in space, and their carrying capacity implied that Soviet missiles armed with nuclear warheads could reach the United States. The resulting political fallout was intense, pressuring President Eisenhower to bolster the US space effort by creating the National Aeronautics and Space Administration.

In April 1961, the Soviets launched *Vostok I*, and Maj. Yuri Gagarin became the first human to orbit Earth. This feat gave further indication of Soviet superiority, exposing a widening gap between Soviet and US rocketry. The ensuing public furor intensified pressure on the government to wrestle space dominance away from the Soviet Union. Otherwise, many thought, the status of the United States would sink to that of a second-class power, with Soviet space prowess exemplifying the superiority of the Communist system.

Newly elected president John F. Kennedy had to do something to leapfrog the Soviet lead in space, something that would restore the country's image to the position of number one in technological leadership. A manned space flight program capable of outdoing the Soviets would require vast expenditures. Such resources were difficult to justify on the basis of what could be learned scientifically. Even Kennedy's science adviser was against manned flight, arguing that most scientific data could be gathered more cheaply using spacecraft that were unmanned or manned only by robots. Nor did it appear that military advantage beyond that obtainable with unmanned satellites and ballistic missiles would follow. Nor was it clear that significant economic benefit would accrue from manned space flight.

Nevertheless, in the public mind the conquest of space was nearly synonymous with manned space flight, and the orbiting cosmonaut made it appear that the Soviet Union was becoming the conqueror of space. The groundswell of popular sentiment became overwhelming and culminated in the president's dramatic response. In May 1961, standing before a joint session of Congress, he announced to the world, "I believe that this nation should commit itself to achieving the goal, before the decade is out, of landing a man on the moon, and returning him safely to Earth." These words set into motion an engineering endeavor of unparalleled proportion.

In physical size, the task's embodiment was gigantic. The design that emerged consisted of a three-stage Saturn V launch vehicle, stacked beneath the Apollo spacecraft. In all, the rocket and spacecraft towered nearly four hundred feet high. More than twice as tall as any US rocket then in existence, its size could not be captured by photographs, for a human figure standing next to it would be indiscernibly small. Only diagrams showing it in proportion to the Statue of Liberty (305 feet) or the Washington Monument (550 feet) could display the size of what was to rise into space. The system's complexity was to be equally daunting. Engineers at NASA identified thousands of tasks to be accomplished, and the Apollo spacecraft would have more than four million parts, arguably exceeding the complexity of any previous manmade artifact.

The president's phrase "before the decade is out" placed a strict deadline on the project, one that would require compressing the development process, which would increase costs and add risks. The president's words implied the even-more-daunting goal of jumping ahead of the Soviet effort and landing astronauts on the moon before they did. Politically, this was to be a continuing challenge, for each time a Soviet launch took place, the world—and the US electorate in particular—would be watching.

President Kennedy's words "returning him safely to Earth" pointed to another challenge. Rocketry to that date had meant primarily military ballistic missiles, which often experienced launch failures during development. Carrying astronauts to the moon and safely back demanded much higher degrees of reliability, while at the same time lifting much larger payloads and dealing with much more difficult problems in communications, stability, and control.

Safety issues would be less of an impediment to the Soviet program. The United States is an open society, and the news media followed the Apollo program closely. Accidents resulting in astronaut fatalities would be viewed as national calamities, with the public demanding to know the causes and looking for someone to blame. Indeed, when three astronauts died in the fire during *Apollo 1* ground testing, the tragedy was broadcast around the world, multiple investigations ensued, and progress toward the moon landing was delayed for many months. In the closed society of the Soviet Union, the government trumpeted space triumphs but shielded accidents and failures from public scrutiny. Secrecy allowed program managers to speed development by taking greater—sometimes unconscionable—risks. Not until decades later, for example, did details of the 1961 Nedelin disaster become known in the West: under pressure from Soviet premier Nikita Khrushchev, Marshall Nedelin rushed to launch a three-stage rocket from the Soviet launch site in central Asia. Troubles developed during the countdown, and the marshall ordered his personnel to climb the rocket's scaffolding to attempt repairs without first emptying the massive fuel tanks. One of the rocket's engines ignited, causing the fuel tanks to explode, engulfing the launch site in a toxic fireball and claiming nearly one hundred lives.

The president had defined a project and deadline that demanded an engineering effort on a colossal scale. Turning his mandate into reality nevertheless followed a process recognizable to engineers, for they already employed it in varied forms to create technology, albeit on a much less ambitious scale than called for by the Apollo program. First has to come conceptual design, examining distinctly different approaches, weighing their pros and cons, and choosing among them. Design then proceeds to more detailed stages, reducing concepts to drawing and specifications, to the dimensions, tolerances, and material compositions needed to build the hardware. As design progresses, it does so against a backdrop of increasing numbers of calculations and simulations and against the testing of concepts and components to ensure that engineers' ideas can be implemented in physical reality. Through testing, inadequacies are found, and flaws uncovered. Designs are changed and refined as engineers gain knowledge of how their creations will behave when put to the actual test. In the Apollo program, as in other engineering endeavors, the cycle of calculations, tests, redesign, and refinement would continue until the astronauts of *Apollo 11* headed toward the moon.

Even before the president's speech, competing concepts for getting to the moon had been circulating around NASA's laboratories. Brainstorming produced even more, some of them bordering on the fanciful, but lengthy discussion and evaluation eliminated those that were clearly unworkable and reduced the candidates to the three concepts depicted in figure 52.

The first, the direct-ascent mode, employed a single launch vehicle powerful enough to boost the entire spacecraft toward the moon. Stages of the rocket would be discarded in flight, and the sections of the spacecraft not needed for the return flight would be abandoned on the moon's surface. Nevertheless, the remaining vehicle would be heavy, carrying rockets sufficiently large to overcome the moon's gravity, to support equipment for the long flight back, and to carry the heat shield needed to bring the astronauts safely through Earth's atmosphere. This option was the first to be abandoned, as it became apparent that the mammoth launch vehicle needed to lift the mission from Earth would require rocket

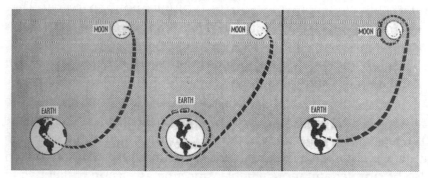

Figure 52. Three lunar landing concepts (left to right): direct ascent, Earth orbit rendezvous, lunar-orbit rendezvous. (Diagram from NASA.)

engines much larger than any that were then under consideration. The development of engines of the unprecedented size needed for the direct-ascent mode in time to meet the president's deadline was out of the question.

A protracted debate raged in 1962 until a choice was made between the two remaining options. For each, launch vehicles of a more reasonable size would suffice. In the Earth-orbit mode, two or more launch vehicles would be required. Following launch of the astronauts and the spacecraft's components into Earth orbit, the astronauts would assemble the spacecraft. A "tanker" vehicle would then carry the fuel needed for the moon mission into Earth orbit, where the astronauts would fuel the assembled spacecraft, and proceed to the moon. At the Marshall Space Flight Center, where the launch vehicles would be designed, Wernher von Braun and his team initially favored this option. It would not only serve to get astronauts to the moon, but the experience would provide know-how in Earth-orbit assembly of the larger spacecraft needed for further space exploration. At that time many within NASA contemplated a very ambitious program culminating with landing on Mars, perhaps by the 1980s.

In the lunar-orbit option, which was finally adopted, a single launch vehicle would lift a three-astronaut spacecraft directly into lunar orbit. A small lunar-landing module carrying two astronauts would detach from the orbiting spacecraft and descend to the moon's surface. After discarding that part of the lander needed only for the descent, the astronauts would launch the ascent module into lunar orbit and rendezvous with the orbiting Apollo spacecraft. They would then abandon the lunar module, and the orbiting spacecraft's rockets would launch the return trip to Earth.

The lunar-orbit concept allowed the use of a lighter spacecraft than the other two modes, both of which required the entire spacecraft to land on the moon and thus to carry much heavier rockets to overcome lunar gravity. Only the lunar-orbit mode permitted the entire expedition to ride on a single launch vehicle of the size then under development. The launching from the moon's surface would also be simpler and more reliable, since the ascension rocket had to be only powerful enough to place the landing module into lunar orbit. The lunar-orbit method also eliminated the complications of assembling and fueling a spacecraft while in Earth orbit. These complications appeared ever larger as the preliminary design studies began to detail the obstacles that would be encountered and particularly the delay they would likely cause. The benefits of lunar orbit overrode its primary drawback: if a rendezvous in Earth orbit failed, the mission could be aborted and the crew returned to Earth, but if it failed in moon orbit, the spacecraft's astronauts would be lost.

Figure 53 shows the hardware implementation of the lunar-orbit concept as it was to emerge as the Saturn V launch vehicle and Apollo spacecraft. The launch vehicle consists of a three-stage rocket. The first two stages lift the Apollo spacecraft into Earth orbit and are then abandoned along with the escape system, included for launch emergencies. The third stage provides thrust to accelerate the spacecraft out of Earth orbit and toward the moon. The spacecraft consists of command and service modules, the command module carrying the astronauts, and the service module supplying oxygen, electrical power, and the rockets needed for maneuvering into lunar orbit and later for the return home. An adapter section carries the lunar module. Once outside Earth's atmosphere and accelerating toward the moon, the astronauts extract the lunar module from the adapter, dock it to the command module, and then abandon the adapter and third-stage rocket.

The mission then proceeds by placing the spacecraft into lunar orbit. Two astronauts then descend to the lunar surface in the lunar module, while the third remains in the command module. The lunar module is equipped with two sets of rockets, one for descent and a second for ascent. Once on the lunar surface the section of the module needed only for descent is detached, thus lessening the weight of the module that must be lifted back into lunar orbit. The lunar and command modules then rendezvous and dock; the two astronauts join the third in the command module; they jettison the lunar module and fire the rockets in the attached service module and head for Earth.

Turning the conceptual design into the detailed reality of hardware and procedures was a massive undertaking. It required close col-

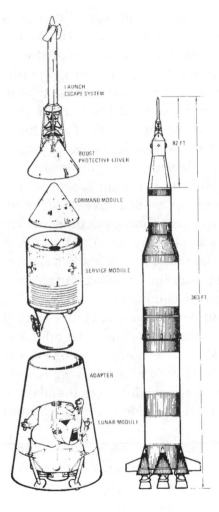

Figure 53. The Apollo spacecraft and the Saturn V launch vehicle. (Diagram from NASA.)

laboration between government and industry to gain the breadth of knowledge and talent needed. The effort was directed and coordinated by NASA, but the agency contracted much of the design and testing to corporations and to their subcontractors. The Apollo spacecraft development was centered at the Manned Spacecraft Center in Houston, but NASA placed the primary contract for building the command and service modules with North American Aviation and for the lunar-landing module with Grumman Aerospace Corporation. Design of the Saturn V launch vehicle was centered at the Marshall Space Flight Center in Alabama under Wernher von Braun's direction, with contracts for the development of rocket engines and the building of Saturn V's three stages going to several different corporations.

The spacecraft's design brought engineers into a new realm, for much of their experience didn't apply to the Apollo command and service modules and even less to the lunar module. Many designers were aeronautical engineers coming fresh from supersonic aircraft projects. But now they had to design a machine to operate in a vacuum, with little or no gravity. Years ago, at least the airplane's form had been agreed upon; it had a fuselage, wings, and a tail. But what form should a lunar module take? Aerodynamic streamlining to cut down air resistance was irrelevant, since the module would operate only in a vacuum, and furthermore, it would experience only the moon's gravity, one-sixth that of Earth. The tacit knowledge the engineers brought from aircraft design and other endeavors was not very helpful, for in space things behaved differently; lubricants, leaks, and the sloshing of fuel in tanks all created challenges.

But how did the designers obtain the knowledge they needed? Certainly they scoured their brains and background for every relevant piece of engineering and made use of the accumulated science of their time. However, many of the questions that they were forced to address had no answers in the theory or experiments of the past, and they had to initiate new research—crash programs to find what they needed to know, and to find it quickly.

While behind the Soviets in accomplishment, the space projects that preceded Apollo were an invaluable source of data and experience. Before the president's announcement, NASA's Mercury program had rocketed a chimpanzee into orbit to study the effects of weightlessness and brought him back safely. Then, less than a month before Kennedy's announcement, the first US astronaut, Alan Shepard, completed a fifteen-minute suborbital flight. By the time John Glenn orbited Earth and the remaining flights of the Mercury project were completed in 1962, the NASA designers had a firmer grip on the human body's response to weightlessness in a space environment, and they also learned a great deal about the design of equipment for use in space. They determined how to shape the space capsule, for example, and how to fabricate the heat shield needed to bring the astronauts through the searing heat created as their spacecraft reentered the atmosphere.

As the design of the Apollo spacecraft and the Saturn V booster got underway, the Gemini project provided further knowledge and data from space. Gemini had created a larger Earth-orbiting space capsule carrying two astronauts. Like Mercury, Gemini used modified military rockets as launch vehicles. With Gemini's ten manned flights, engineers developed techniques for locating, rendezvousing, and docking spacecraft, and astronauts learned to work outside the spacecraft as would be required in lunar-orbit rendezvous. The program also provided assurance of human ability to function in a weightless space environment for the two weeks or longer that might be needed to get to the moon and back. The orbiters provided a test-bed for improvements in spacecraft design: the substitution of fuel cells for batteries, for example. And during manned flights in 1965 and 1966, Gemini increased engineers' ability to design life support, communications, and other onboard systems that would be needed to take astronauts to the moon.

Earth-orbit flights, however, could not provide all of the data critical to the design of the spacecraft. The moon's surface characteristics, its ability to hold the weight of the lunar module, radiation, and micrometeorite dangers all required evaluation. For this, NASA launched unmanned probes deeper into space. A lunar orbiter obtained close-up pictures of the moon's surface in 1966, and in 1967 a Surveyor satellite made a soft landing on the moon and determined that the crust was sturdy enough to support the lunar module's weight.

With what knowledge they had, engineers began sketching and doing preliminary calculations. They contemplated what shape the lunar module should take, how it would be laid out, how large the rockets had to be for descent and ascent, and how much fuel had to be carried. They sought solutions to problems big and small: How would the astronauts see the moon's surface as they approached landing? How would they get out of and back into the lander? What should the support structure look like? Should it be retractable or fixed? And how many legs should it have?

From their drawings, numbers, and specifications, the engineers first embodied their designs in scale models and full-sized mock-ups. Initially made of wood simply to specify the size, shape, and layout of the lunar module without being in any way functional, the mock-ups then became progressively more sophisticated. These were frequently referred to as "boilerplate" within the Apollo program. Later mock-ups were constructed from metal and properly weighted, and some were used to test the Saturn launch vehicles, since they approached the size, weight, and shape of the final product. Astronauts also participated, climbing aboard the mock-ups, making suggestions, and providing user feedback early in the design process. The size and shape of entry hatches, the type and placement of controls, and the windows' size and location all evolved through feedback from those who would occupy the landing module. The astronauts also collaborated in determining the sequence of procedures that the structure filled with pipes, tanks, and computers must negotiate: landing, ascent, rendezvous, and docking.

As in all spacecraft, reducing weight to a minimum was a necessity. For excess weight meant that the lunar lander could carry less fuel and could fire its rockets for shorter periods of time, thereby reducing safety margins in the event that problems required the mission to be redirected or aborted. The engineers redesigned components to weigh less, shaved metal from whatever they could, and reduced the thickness of pipes and tanks to limits below which safety would become imperiled. They designed the ladder from which astronauts would step to the lunar surface strong enough to support them only under the force of the moon's gravity. On Earth, it would collapse under their weight.

But how could they be confident that their creation would work? Theory and calculations were not enough, and although ground testing of mock-ups and prototypes was a necessary first step, the weightlessness and vacuum of space were sure to bring surprises that lay outside the store of scientific knowledge from which they worked. They needed to determine how both the machines and their occupants would behave in the vacuum of space and how the changes from the strong pull of acceleration during launch to the weightlessness of space would affect the performance of what they designed. To understand the effects of the absence of gravity, they placed mock-ups and prototypes in NASA aircraft. By flying carefully programmed parabolic flight paths, the pilots could create several minutes of free-fall conditions, during which astronauts and test engineers could examine the behavior of module components. Likewise, inverted parabolic flight patterns exposed equipment and crew to the strong forces similar to that they would experience during liftoff. The vacuum of space created other challenges: for example, the lunar module's rocket engines had to be able to start and stop in a vacuum. Thus NASA built a gigantic vacuum chamber in the New Mexico desert to test the module's engines before they attempted to operate them in space.

Also critical to the spacecraft's success were the simulators that the engineers designed and built. These they used to train the astronauts on the ground and give them practice for normal flight and even more for dealing with unplanned occurrences, failed equipment, and other emergencies. Simulators were invaluable in testing mission rules, procedures, and techniques in order to gain confidence that what had been worked out on paper would also work in practice.

Parallel to the Apollo spacecraft development, engineering of the Saturn launch vehicle pushed ahead. Following a scaling-up program, the engineers produced a series of the three successively larger and more powerful launch vehicles, shown in figure 54. Each made use of elements of the preceding vehicle but also required innovative technology and had to overcome significant challenges.

The Saturn I and IB were two-stage rockets. The Saturn I utilized two existing engines. The first stage, the H-1, relied heavily on military missile technology, using liquid oxygen with conventional kerosene-based fuel. The second stage, the RL10 rocket engine, was more innovative, combing liquid hydrogen and oxygen and achieving a higher performance-to-weight ratio. Each stage developed the multiengine concept, which allowed the launch vehicle stage to be controlled more easily than with a single, larger engine. Moreover, even if one or possibly two of the engines failed during a manned mission, the multiple engines in each stage increased the likelihood that the astronauts could be brought safely back.

The Saturn IB was an interim vehicle. It would not have sufficient thrust to carry Apollo to the moon, but its function was vital just the same. The IB was to allow Earth-orbit testing of the Apollo and Saturn hardware while the Saturn V was still under development. The tests were to be performed on the command and service modules, on the lunar excursion module, and on the rocket that would be employed in the second and third stages of the Saturn V.

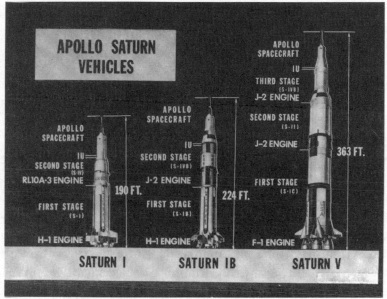

Figure 54. Rocket engine designations for the
Saturn I, IB, and V launch vehicles. (Diagram from NASA.)

The IB had an upgraded version with the same first stage as the
Saturn I. However the newly developed J-2 engine powered the
second stage with liquid hydrogen. Five J-2s were to constitute
Saturn V's second stage, and a sixth was employed to power the third
stage for Apollo flights. Engineers used IB tests to assure the reliability of the J-2's new hydrogen technology and to demonstrate that
the engine would start, shut down, and restart in space. They first
flight-tested the Saturn IB by lifting a mock-up of an early Apollo
spacecraft design into Earth orbit and also used it to launch
unmanned satellites. The IB would later carry manned Apollo spacecraft and lunar module into Earth orbit for testing.

The larger Saturn V included three stages, with the third-stage
rocket powerful enough to accelerate the Apollo spacecraft out of
Earth orbit and on its way to the moon. Whereas the J-2 engines
used in the second and third stages had undergone earlier Saturn IB
testing, the first stage was completely new. Powered by five newly

developed F-1 engines, it generated much greater thrust than the Saturn I or IB. The F-1 used conventional propellants, but it was awesome in size and thrust. NASA conceived the engine to leapfrog the Soviets' ability to lift larger payloads than the United States. The engine was a gamble, and the leap forward did indeed cause difficulties. During ground testing, the F-1's combustion became unstable and generated explosive forces so strong that they destroyed the engine. Engineers spent months attempting to understand the nature of the instabilities and to design around them. The instabilities destroyed three test engines before crash research efforts and multiple component redesigns finally put the problem to rest.

Rocket engines, spacecraft hardware, and much more went through the cycles of design, copious analysis, relentless testing, and redesign that epitomize the engineering process. Within the Apollo program, as within many development projects, philosophies differed as to the thoroughness of subsystems and component testing that should take place before attempting to operate a system as a whole. Such differences came to the front as NASA determined at what point to proceed from ground to flight testing, and from unmanned to manned flight testing. Flight testing was essential, for at some point the largest uncertainties that remained became those that could be resolved only in space. Yet each flight test would be very expensive and risky as well when astronauts were aboard.

Conservatism might call for a step-at-a-time approach—for example, testing each stage of the new Saturn rockets in space and eliminating its problems before adding the next. But if the incremental knowledge gained from each test is small, then some may ask if the expense can be justified. And even if there is nothing new in the rocket's design, astronauts are exposed nonetheless to substantial risk each time they venture into space. Conversely, if testing is too bold, if too many components or procedures are new, chances for catastrophic failure multiply, and risks become intolerable, even when weighed against the greater knowledge to be gained. Initially, the NASA engineers, particularly the designers of the Saturn launch vehicles, favored a conservative step-by-step approach. At one point they planned sixteen Apollo flights to precede the lunar landing. But to meet the president's deadline and to live within ever-tightening budgets, pressure grew to follow the bolder, if riskier, strategy of "all-up," which condensed the testing into substantially fewer flights.

Apollo flight testing began using the Saturn I and then the IB to gain confidence in the hydrogen-fueled J-2 engine. The flights also lifted unmanned prototypes of the spacecraft into Earth orbit to ensure that they were rugged enough to survive the shock and vibration of launch and that they would perform in the gravity-free vacuum of space. Then as preparations for Apollo's first manned space flight were under way, a disastrous setback struck. Fire broke out in the Apollo spacecraft, as noted earlier, killing Gus Grissom, Edward White, and Roger Chaffee, the three astronauts who were in the craft atop a Saturn IB launch vehicle. They perished while conducting ground tests in preparation for their Earth-orbit flight.

The tragedy set back the program for many months while investigations took place, pointing out safety lapses, design flaws, and manufacturing deficiencies. But with design corrections and added emphasis on reliability and quality assurance, the Apollo program slowly regained its footing. In memory of the astronauts, the destroyed spacecraft was immortalized as *Apollo 1*. Three unmanned flights that preceded the disaster were to be designated Apollo 1 through 3; instead, they remained nameless.

Three months after the fire, tragedy struck Apollo's Soviet competition as well. Driven by politics, the Soviets undertook an ambitious mission to demonstrate their lead and overcome their failure to show docking capabilities on three previous flights. Early in the flight, however, the spacecraft's solar panels failed catastrophically, causing damage to the communications and control systems. The cosmonaut attempted to descend under manual control, but the craft overheated as it plunged into the atmosphere. The results were parachute failure, explosion of the landing rockets on impact, and the pilot's death.

These tragedies notwithstanding, the Apollo program resumed flight testing on November 9, 1967. In order to accelerate launch vehicle and spacecraft development, NASA pursued its "all-up" policy by testing all three stages of the Saturn V launch vehicle simultaneously. Many still had reservations about the policy, for the Saturn V first stage consisted of five of the mammoth F-1 engines, the new design that had never been tested in flight. Prudence seemed to indicate that the first stage should be tested separately before the second and third stages were stacked upon it. Moreover, whereas the J-2 engines of the upper stages had been flight-tested one at a time, Saturn V's second stage consisted of the untested arrangement of placing five J-2s in tandem. Program management overrode the objections, however, and following many delays of the launch to fix problems and correct flaws, the Saturn V lifted off in October 1967 and performed admirably in the flight designated as *Apollo 4*.

The Saturn V's second flight, designated *Apollo 6*, took place on April 4, 1968. It was to provide further engine performance validation. Equally important, it tested the navigation systems, the third-stage restart capability, and the Apollo command module's heat shield. Whereas the first all-up flight of the Saturn V had been a resounding success, troubles developed on *Apollo 6*. Two minutes after launch the first stage began to oscillate, lurching forward and backward like a pogo stick for ten seconds, to an extent threatening the rocket's destruction. Similar oscillations had occurred in guided missiles and in *Apollo 4*, but they had been too mild to be worrisome. More problems ensued. Two of the second stage's five rocket engines shutdown prematurely, and only by firing the remaining engines for a longer period than planned did the flight proceed. Finally, the test to restart the third-stage engine, as would be required for a moon flight, failed. Fortunately, the Apollo service module engine allowed the reentry test of the heat shield to be performed successfully, and the unmanned spacecraft returned to Earth intact.

Between the two Saturn V flights, parallel work continued. On January 22, 1968, the Saturn IB that was to have been used in *Apollo 1* became the launch vehicle for *Apollo 5*, an unmanned Earth-orbit flight that tested parts of the lunar module, particularly the capabilities of small rocket motors to be used for moon landing. Then finally, on October 11, 1968, a second Saturn IB launched the first manned flight, *Apollo 7*, placing three astronauts in Earth orbit to test the spacecraft's capabilities.

Arguably the most daring of the test flights was *Apollo 8*. By then the problems that had plagued Saturn V's *Apollo 6* flight had been diagnosed, and design alterations were made that garnered enough confidence for NASA to bypass further unmanned flights and schedule the Saturn V to launch astronauts into space. Engineers originally planned for a manned Earth-orbit flight to validate the compatibility of the Apollo spacecraft with the Saturn V launch vehicle. Intelligence sources, however, indicated that the Soviets were planning a lunar flight, and the United States had to do something dramatic to prevent the appearance of falling further behind in space. The navigational accuracy required was unprecedented, and other challenges abounded as well. Nevertheless, with all-up strategy, the first manned mission of Saturn V carried astronauts into orbit around the moon. On Christmas 1968 NASA broadcasted dramatic close-up pictures of the moon's surface and of distant Earth, along with the astronauts' holiday greetings, maximizing the event's political impact.

Following *Apollo 8*, engineers required two less dramatic missions to prepare for the moon landing. *Apollo 9*, the last Saturn IB flight, lifted the spacecraft into Earth orbit so that astronauts could practice flying the lunar lander, maneuver it, and dock it with the Apollo command module. A Saturn V then launched *Apollo 10*: a full dress rehearsal in which astronauts orbited the moon and tested everything short of the actual descent of the lunar module to the moon's surface. Some argued for landing on the moon then, but the tests to modify the landing software to accommodate variations in the moon's gravitational field first had to be performed and evaluated.

The climactic flight, *Apollo 11*, was launched on July 16, 1969, and landed the astronauts on the moon within the decade as President Kennedy had sought. The flight was a dramatic achievement of twentieth-century technology: it brought to fruition a nine-year progression of project definition, design innovation, computation, scaling-up, testing, and evaluation that exemplifies the core processes through which engineers bring about technological advance.

❀   ❀   ❀

Well before *Apollo 11* was launched, public support for the massive expenditures required for manned space exploration had begun to wane. Earlier visions of a far more ambitious program to land astronauts on Mars soon gave way to the more earthly concerns of guerrilla war in Southeast Asia and pressing social and economic issues at home. In the ensuing decades the more limited space shuttle program has had difficulty maintaining adequate support, and the astronauts who tragically perished in the *Challenger* and *Columbia* shuttle disasters have indelibly marked in the public's mind that seemingly routine manned missions in space entail high risks.

Across many fronts, technology and the science and engineering behind it have advanced greatly in the decades since *Apollo 11* landed on the moon. The rise of integrated circuits, fiber optics, and other microelectronic innovations has led to an explosion in new technology: the Internet, cellular telephones, medical diagnostic imaging, satellite communications, and more. Likewise, advances in biotechnology are emerging with immense potential for the future. Arguably, these technological advances account for far more of an impact on society and the lives of its citizens than manned space flight and the lunar landings have had.

Whereas these more recent advances are often closely attuned to health, the economy, and other day-to-day concerns of the citizenry, they do not hold the drama or the spectacular physical embodiment of technology found in the race to reach the moon. Only exceptional circumstances bring a society together in providing the resources required to succeed in an undertaking of such a magnitude—particularly one that fulfilled no pressing economic necessity or military need. But it has happened before. The building of the pyramids, of great cathedrals, and of skyscrapers was financed by societal concentration of religious fervor, civic pride, commercial hubris, or national competition. One need only look at the succession of pyramids or at the Gothic spires whose heights climbed with the pride of competing cities. In more modern times the New York skyline was defined by the competition of Wall Street financiers and corporate chieftains to build ever-taller skyscrapers, culminating in 1930 with the Empire State Building. Although dampened by the

World Trade Center disaster, a worldwide competition for building taller buildings persists even today. Thus in the broadest sense, the cold war space competition that resulted in the US lunar landing was not unique in technological history.

It's not surprising that the advances in rocketry that led to the *Apollo 11* flight to the moon should enshrine the catchphrase that has come to encapsulate the twentieth century's technological prowess. But was the creation of Apollo's rocketry science? Yes, but only in part.

When present-day engineers set pencil to paper—or electronic pen to computer screen—and begin sketching a design, they do so not only with the tacit knowledge gained from experience but also with a thorough grounding in the physical sciences. This knowledge provides them with opportunities that their predecessors didn't have. It also reins in their imaginations, preventing them from expending effort on fanciful ideas that clearly would violate well-understood scientific principles. As engineers draw, theorize, and calculate—as they collaborate with architects, industrial designers, and others—issues inevitably arise, and questions appear for which the solutions are not apparent. Some are social, revolving about clients' conflicting needs or customer preferences, but many are scientific, exposing shortcomings in the designers' knowledge. They then must consult handbooks, search relevant databases, or perform more exacting calculations before their design can progress. The thorniest problems often come down to the fact that the engineers' needs have outrun the information that the current state of science is able to provide. Sometimes technological innovators may be able to redesign around their gaps in knowledge; otherwise they must propose experiments, perform tests, or originate crash programs in scientific research. They target such programs, however, to provide their project with the information needed to proceed with the design within an allowable time frame. Their motivation differs sharply from the quest of pure science for a more fundamental understanding of the nature of the physical universe.

That's not to say that new technology must originate from engineers' pulling together of what already exists in ingenious ways to meet their clients' needs. Often fundamental scientific discoveries do lead to new technology, and dramatically so. Thus, whereas the Apollo program resulted from a political mandate and built on the technology then available, nuclear energy resulted directly from the discovery of the neutron and of fission. Likewise, the understanding of the quantum physics of solids led to the transistor's invention. Nevertheless, following discovery and the realization of its technological potential, there inevitably follows an extended period of engineering, as scientific discovery is melded with human need through design, analysis, testing, and evaluation.

The knowledge with which Apollo engineers approached their tasks and their quests for scientific findings that were triggered by needs to answer design questions—the "rocket science" they employed—differed subtly but significantly from the common conception of engineering as applied science. Overall the project presented a problem to be solved using the science that was then available. The science and technology had to be "off the shelf," or at a state where it could be brought out of the research lab very rapidly. It did not follow the converse and less frequent path, in which a physicist or a chemist makes a fundamental discovery and then parlays it to engineers to create a useful product. Even in situations that more closely fit such a scenario—the applications of lasers to eye surgery or microwaves to cooking, for example—the creation of a useful technology has required combining many other things with scientific discovery to achieve the task at hand.

From cosmology to molecular biology, we value immeasurably our deepening scientific understanding of the natural universe, gauging our society's intellectual climate by the strength of our discoveries and the pace at which we accumulate knowledge. We likewise treasure the insights gained from the great works of the humanistic tradition, judging our culture's vitality by the talents of our musicians, actors, and dancers and by the creations of our writers, poets, painters, composers, and choreographers. Indeed, we may rightly place the basic sciences and fine arts at the pinnacle of human endeavors. The practical arts, however, should occupy a status that parallels those lofty disciplines. The work of architects and industrial designers does much to enrich our physical environment—ranging from the tallest buildings to the smallest consumer products—creating artistic delight where ugliness might otherwise reign and bringing convenience where frustration would likely hold sway. And while engineers' labors are not focused on the pursuit of enduring truth or wisdom, the technology that they create plays a large part in satisfying humanity's more immediate material needs and aspirations.

Indeed, without the material assets that technology provides, humanity could not appreciate higher cultural attainments, certainly not more than an aristocratic few. In *The Ancient Engineers*, L. Sprague de Camp puts it better than most: "Civilization, as we know it today, owes its existence to the engineers. These are the men who, down the long centuries, have learned to exploit the properties of matter and the sources of power for the benefit of mankind. By an organized, rational effort to use the material world around them, engineers devised the myriad comforts and conveniences that mark the difference between our lives and those of our forefathers thousands of years ago."[1]

Those who look beyond the material benefits brought by engineering to the tools, techniques, and thought processes that have gone into their creation encounter a most fascinating pursuit: a vocation driven by an exuberance to conceive new technology, to build it, and to make it work. They discover a profession that utilizes modern scientific methods and discoveries but is grounded in a heritage whose roots date back long before the scientific or industrial revolutions to the earliest civilizations.

Technology, and the engineering that creates it, offers no panacea. It cannot overcome greed, bigotry, or the will to power. The ever-expanding cornucopia of technological possibilities that lie before us, however, can open doors and create opportunities. Knowledge and wisdom are vital in choosing and molding those opportunities to human benefit, to bringing hope to all the world's peoples that their material needs will be met through technologies that are sustainable over the long term, that conserve natural resources and protect the environment. Engineering must participate heavily in those choices but cannot dictate them, for the choices are inherently social as well as technological. Wise technological decisions require an ongoing dialogue, deepening engineers' understanding of the society that they serve and increasing the public's comprehension of technology and of the human endeavor through which it is created. My hope is that these pages will provide some measure of appreciation for engineering's past as well as an understanding of its present and its relationship to science and to society. Perhaps they may thereby serve in some small way as prologue for those who will forge the future.

Notes

CHAPTER 1

1. For the importance of failure in engineering, see Henry Petroski, *To Engineer Is Human: The Role of Failure in Successful Design* (New York: St. Martin's Press, 1985).

CHAPTER 2

1. For additional specifics on nineteenth-century wagon construction, see George Sturt, *The Wheelwright's Shop* (Cambridge: Cambridge University Press, 1923).

2. See, for example, George Basalla, *The Evolution of Technology* (Cambridge: Cambridge University Press, 1988); M. J. French, *Invention and Evolution: Design in Nature and Engineering* (Cambridge: Cambridge University Press, 1988).

3. Oliver Wendell Holmes, "The Deacon's Masterpiece, or, The Wonderful 'One-Hoss Shay': A Logical Story," in *Yale Book of American Verse*, ed. Thomas R. Lounsbury (New Haven, CT: Yale University Press, 1912).

CHAPTER 3

1. John Fitchen, *The Construction of Gothic Cathedrals* (London: Oxford University Press, 1961).

2. Jacques Heyman, *The Stone Skeleton* (Cambridge: Cambridge University Press, 1995); Robert Mark, *Experiments in Gothic Structure* (Cambridge, MA: MIT Press, 1982).

CHAPTER 4

1. Arnold of Bonneval, *Sancti Bernardi abbatis Clarae-Vallensis vita et res gestae libris septum comprehensae* 2.5; trans. David Luckhurst, *Monastic Watermills: A Study of the Mills within English Monastic Precincts* (London: Society for the Protection of Ancient Buildings, n.d.), p. 6.
2. See Tom D. Crouch, *A Dream of Wings* (Washington, DC: Smithsonian Institution Press, 1989), p. 27.

CHAPTER 5

1. See, for example, Basil Greenhill, *The Evolution of the Wooden Ship* (New York: Facts on File, 1988).
2. Eugene S. Ferguson, *Engineering and the Mind's Eye* (Cambridge, MA: MIT Press, 1993).

CHAPTER 6

1. For more on Galileo and machinery, see Arnold Pacey, *Maze of Ingenuity: Ideas and Idealism in the Development of Technology* (Cambridge, MA: MIT Press, 1992), chap. 3.
2. See Brett D. Steele, "Muskets and Pendulums: Benjamin Robins, Leonard Euler, and the Ballistics Revolution," in *Technology and the West*, ed. T. S. Reynolds and S. H. Cutcliffe (Chicago: University of Chicago Press, 1997), pp. 145–79.

CHAPTER 7

1. Claude Mosse, *The Ancient World at Work* (London: Chatto & Windus, 1969), p. 79.

2. Adam Smith, *The Wealth of Nations* (Amherst, NY: Prometheus Books, 1991), p. 13.

3. Frederick W. Taylor, *The Principles of Scientific Management* (New York: Norton, 1967).

CHAPTER 9

1. Walter G. Vincenti, *What Engineers Know and How They Know It: Analytical Studies from Aeronautical History* (Baltimore: Johns Hopkins University Press, 1990).

CHAPTER 10

1. Bad and good designs are compared in Donald A. Norman, *The Design of Everyday Things* (New York: Basic Books, 1988).

2. An example of this is the disastrous balcony collapse at the Kansas City Regency Hotel in 1981; see Henry Petroski, *To Engineer Is Human: The Role of Failure in Successful Design* (New York: St. Martin's Press, 1985), chap. 8.

CHAPTER 11

1. L. Sprague De Camp, *The Ancient Engineers* (New York: Ballantine Books, 1963), p. 1.

Select Bibliography

CHAPTER 1

Armytage, W. H. G. *A Social History of Engineering*. Cambridge, MA: MIT Press, 1961.

Cardwell, Donald [S. L.]. *The Norton History of Technology*. New York: Norton, 1995.

De Camp, L. Sprague. *The Ancient Engineers*. New York: Ballantine Books, 1963.

Edwards, I. E. S. *The Pyramids of Egypt*. New York: Penguin, 1985.

Kirby, Richard Shelton, Sidney Withington, Arthur Burr Darling, and Frederick Gridley Kilgour. *Engineering in History*. New York: Dover, 1990.

Lepre, J. P. *The Egyptian Pyramids*. Jefferson, NC: McFarland, 1990.

McClellan, James E., and Harold Dorn. *Science and Technology in World History*. Baltimore: Johns Hopkins University Press, 1999.

Pacey, Arnold. *Maze of Ingenuity: Ideas and Idealism in the Development of Technology*. 2nd ed. Cambridge, MA: MIT Press, 1992.

———. *Technology in World Civilization*. Cambridge, MA: MIT Press, 1992.

Petroski, Henry. *To Engineer Is Human: The Role of Failure in Successful Design*. New York: St. Martin's Press, 1985.

Price, Derek de Solla. *Science since Babylon*. 2nd ed. New Haven, CT: Yale University Press, 1975.

Singer, C., E. J. Holmyard, and A. R. Hall. *A History of Technology*. Vols 1–5. Oxford: Oxford University Press, 1954–58.

CHAPTER 2

Bailey, Jocelyn. *The Village Wheelwright and Carpenter*. Princes Risborough, Buckinghamshire, UK: Shire, 1998.

Basalla, George. *The Evolution of Technology*. Cambridge: Cambridge University Press, 1988

French, M. J. *Invention and Evolution: Design in Nature and Engineering*. Cambridge: Cambridge University Press, 1988.

Jones, J. Christopher. *Design Methods*. New York: Wiley, 1980.

Petroski, Henry. *The Evolution of Useful Things*. New York: Knopf, 1993.

Seymour, John. *The Forgotten Arts*. London: Dorling Kindersley, 1984.

Singer, C., E. J. Holmyard, and A. R. Hall. *A History of Technology*. Vols 1–5. Oxford: Oxford University Press, 1954–58.

Strut, George. *The Wheelwright's Shop*. London: Cambridge University Press, 1923.

CHAPTER 3

Coldstream, Nicola. *Masons and Sculptors*. London: British Museum Press, 1995.

———. *Medieval Architecture*. Oxford: Oxford University Press, 2002.

Fitchen, John. *The Construction of Gothic Cathedrals*. London: Oxford University Press, 1961.

———. *Building Construction before Mechanization*. Cambridge, MA: MIT Press, 1986.

Gimpel, Jean. *The Cathedral Builders*. London: Random House, 1983.

Heyman, Jacques. *The Stone Skeleton*. Cambridge: Cambridge University Press, 1995.

———. *The Science of Structural Engineering*. London: Imperial College Press, 1999.

Mark, Robert. *Experiments in Gothic Structure*. Cambridge, MA: MIT Press, 1982.

Merlet, Rene. *The Cathedral of Chartres*. Paris: Henri Laurens, 1939.

Swaan, Wim. *The Gothic Cathedral*. New York: Park Lane, 1984.

CHAPTER 4

Adam, Jean-Pierre. *Roman Building: Materials and Construction*. London: B. T. Batsford, 1994.

Beck, Willi, and Dieter Planck. *Der Limes in Südwestdeutschland*. Stuttgart: Konrad Theiss, 1980.

Cipolla, Carlo M. *Guns, Sails, and Empires*. Manhattan, KS: Sunflower University Press, 1985.

Crouch, Tom D. *A Dream of Wings*. Washington, DC: Smithsonian Institution Press, 1989.

Dickinson, H. W. *James Watt, Craftsman and Engineer*. London: Cambridge University Press, 1935.

Freudenthal, Elsbeth E. *Flight into History*. Norman: University of Oklahoma Press, 1949.

Gimpel, Jean. *The Medieval Machine*. New York: Penguin, 1976.

Hills, Richard L. *Power from Steam*. Cambridge: Cambridge University Press, 1989.

Hodge, A. Trevor. *Roman Aqueducts and Water Supplies*. 2nd ed. London: Duckworth, 2002.

Law, John. "On the Social Explanation of Technical Change: The Case of the Portuguese Maritime Expansion." In *Technology and the West*, edited by T. S. Reynolds and S. H. Cutcliffe, pp. 119–44. Chicago: University of Chicago Press, 1997.

Prestage, Edgar. *The Portuguese Pioneers*. London: A. & C. Black, 1933.

Reynolds, Terry S. *Stronger Than a Hundred Men*. Baltimore: Johns Hopkins University Press, 1983.

Singer, C., E. J. Holmyard, and A. R. Hall. *A History of Technology*. Vols 1–5. Oxford: Oxford University Press, 1954–58.

Ure, John. *Prince Henry the Navigator*. London: Constable, 1977.

CHAPTER 5

Baynes, Ken, and Francis Pugh. *The Art of the Engineer*. Woodstock, NY: Overlook Press, 1981.

Booker, Peter Jeffrey. *A History of Engineering Drawing*. London: Northgate, 1979.

Cianchi, Marco. *Leonardo's Machines*. Florence: Becocci Editore, 1984.

Ferguson, Eugene S. *Engineering and the Mind's Eye*. Cambridge, MA: MIT Press, 1993.

Gille, Bertrand. *Engineers of the Renaissance*. Cambridge, MA: MIT Press, 1966.

Hindle, Brooke. *Emulation and Invention*. New York: Norton, 1981.

Pedretti, Carlo. *Leonardo, Architect*. New York: Rizzoli, 1985.

White, John. *The Birth and Rebirth of Pictorial Space*. Cambridge, MA: Harvard University Press, 1986.

Zammattio, Carlo, Augusto Marinoni, and Anna Maria Brizio. *Leonardo the Scientist*. New York: McGraw-Hill, 1980.

CHAPTER 6

Corlett, Ewan. *The Iron Ship*. Bradford-on-Avon, UK: Moonraker Press, 1975.

Drake, Stillman. *Galileo at Work*. Chicago: University of Chicago Press, 1978.

———. *Galileo: Pioneer Scientist*. Toronto: University of Toronto Press, 1990.

Galilei, Galileo. *Two New Sciences*. Translation & notes by S. Drake. 2nd ed. Toronto: Wall & Thompson, 1989.

Goldstein, Thomas. *Dawn of Modern Science*. Boston: Houghton Mifflin, 1980.

Griffiths, Denis, Andrew Lambert, and Fred Walker. *Brunel's Ships*. London: Chatham, 1999.

Machamer, Peter, ed. *The Cambridge Companion to Galileo*. Cambridge: Cambridge University Press, 1998.

McClellan, James E., and Harold Dorn. *Science and Technology in World History*. Baltimore: Johns Hopkins University Press, 1999.

Pacey, Arnold. *Maze of Ingenuity: Ideas and Idealism in the Development of Technology*. 2nd ed. Cambridge, MA: MIT Press, 1992.

Rolt, L. T. C. *Isambard Kingdom Brunel*. London: Longmans, Green, 1957.

Sanders, Michael S. *The Yard: Building a Destroyer at the Bath Iron Works*. New York: HarperCollins, 1999.

Shapin, Steven. *The Scientific Revolution*. Chicago: University of Chicago Press, 1996.

Skeppton, A. W., ed. *John Smeaton, FRS*. London: Thomas Telford, 1981.

Smith, Dennis. "The Use of Models in Nineteenth Century British Suspension Bridge Design." In *History of Technology*, vol. 2, edited by A. R. Hall and N. Smith, pp. 169–214. London: Mansell, 1977.

Steele, Brett D. "Muskets and Pendulums: Benjamin Robins, Leonard Euler, and the Ballistics Revolution." In *Technology and the West*, edited by T. S. Reynolds and S. H. Cutcliffe, pp. 145–79. Chicago: University of Chicago Press, 1997.

Timoshenko, Stephen P. *History of Strength of Materials*. New York: Dover, 1983.

CHAPTER 7

Alder, Ken. *Engineering the Revolution*. Princeton, NJ: Princeton University Press, 1997.

Bernal, J. D. *Science in History*. Vol. 2, *The Scientific and Industrial Revolutions*. Cambridge, MA: MIT Press, 1969.

Buderi, Robert. *The Invention That Changed the World*. New York: Simon & Schuster, 1996.

Cowan, Ruth Schwartz. *A Social History of American Technology*. New York: Oxford University Press, 1997.

Hirschhorn, Larry. *Beyond Mechanization: Work and Technology in a Postindustrial Age*. Cambridge, MA: MIT Press, 1986

Hounshell, David A. *From the American System to Mass Production, 1800–1932*. Baltimore: Johns Hopkins University Press, 1984.

McKendrick, Neil. "Josiah Wedgewood and Factory Discipline." In *The Rise of Capitalism*, edited by D. S. Landes, pp. 68–81. New York: Macmillan, 1966.

Mosse, Claude. *The Ancient World at Work*. London: Chatto & Windus, 1969.

Nevins, Allan, and Frank Ernest Hill. *Ford: Expansion and Challenge, 1915–1933*. New York: Charles Scribner's Sons, 1957.

Rosenberg, Nathan. *Inside the Black Box: Technology and Economics*. Cambridge: Cambridge University Press, 1982.

Smith, Merritt Roe. *Harpers Ferry Armory and the New Technology: The Challenge of Change*. Ithaca, NY: Cornell University Press, 1979.

Woodbury, Robert S. *Studies in the History of Machine Tools*. Cambridge: MIT Press, 1972.

CHAPTER 8

Buderi, Robert. *Engines of Tomorrow*. New York: Simon & Schuster, 2000.

Cardwell, D[onald] S. L. *From Watt to Clausius*. Ithaca, NY: Cornell University Press, 1971.

Coe, Lewis. *The Telegraph: A History of Morse's Invention and Its Predecessors in the United States*. Jefferson, NC: McFarland, 1993.

Freeman, Christopher. *The Economics of Industrial Innovation*. 2nd ed. Cambridge, MA: MIT Press, 1982.

Grayson, Lawrence P. *The Making of an Engineer: An Illustrated History of Engineering Education in the United States and Canada*. New York: Wiley, 1993.

Hecht, Jeff. *Laser Pioneers*. San Diego: Academic Press, 1992.

Hoddeson, Lillian. "The Emergence of Basic Research in the Bell Telephone System, 1875–1915." In *Technology & the West*, edited by T. S. Reynolds and S. H. Cutcliffe, pp. 331–63. Chicago: University of Chicago Press, 1997.

Israel, Paul. *From Machine Shop to Industrial Laboratory*. Baltimore: Johns Hopkins University Press, 1992.

Kranzberg, Melvin, and Caroll W. Pursell Jr., eds. *Technology in Western Civilization*. Vols. 1 and 2. New York: Oxford University Press, 1967.

Layton, Edwin. "Mirror-Image Twins: The Communities of Science and Technology in 19th-Century America." In *The Engineer in America*, edited by T. S. Reynolds, pp. 229–47. Chicago: University of Chicago Press, 1991.

Marcus, Alan I., and Howard P. Spegal. *Technology in America: A Brief History*. New York: Harcourt Brace Jovanovich, 1989.

Millard, Andre. *Edison and the Business of Innovation*. Baltimore: Johns Hopkins University Press, 1990.

Reynolds, Terry S. "Defining Professional Boundaries: Chemical Engineering in the Early 20th Century." In *The Engineer in America*, edited by T. S. Reynolds, pp. 343–65. Chicago: University of Chicago Press, 1991.

Smith, Crosbie. *The Science of Energy*. Chicago: University of Chicago Press, 1998.

Standage, Tom. *The Victorian Internet*. New York: Walker, 1998.

Townes, Charles H. *How the Laser Happened*. Oxford: Oxford University Press, 1999.

Wise, George. "A New Role for Professional Scientists in Industry: Industrial

Research at General Electric, 1900–1916." In *Technology and American History*, edited by S. H. Cuttcliffe and T. S. Reynolds, pp. 217–38. Chicago: University of Chicago Press, 1997

CHAPTER 9

Aronstein, David C., Michael J. Hirschberg, and Albert C. Piccirillo. *Advanced Tactical Fighter to F-22 Raptor*. Reston, VA: American Institute of Aeronautics and Astronautics, 1998.

Birtles, Philip. *Boeing 777*. London: Airlife, 1998.

Case Study in Aircraft Design: The Boeing 727. American Institute of Aeronautics and Astronautics Professional Study Series, 1978.

Constant, Edward W. II. *The Origins of the Turbojet Revolution*. Baltimore: Johns Hopkins University Press, 1980.

Droste, Carl S., and James E. Walker. *General Dynamics Case Study on the F-16 Fly-by-Wire Flight Control System*. AIAA Professional Study Series, n.d.

Misa, Thomas J. "Military Needs, Commercial Realities, and the Development of the Transistor, 1948–1958." In *Military Enterprise and Technological Change*, edited by Merritt Roe Smith, pp. 253–87. Cambridge, MA: MIT Press, 1987.

Norman, Donald A. *The Invisible Computer*. Cambridge, MA: MIT Press, 1998.

Petroski, Henry. *Invention by Design: How Engineers Get from Thought to Thing*. Cambridge, MA: Harvard University Press, 1996.

Rich, Ben R., and Leo Janos. *Skunk Works*. Boston: Little, Brown, 1994.

Rogers, Eugene. *Flying High*. New York: Atlantic Monthly Press, 1996.

Rolt, L. T. C. *Victorian Engineering*. London: Penguin, 1970.

Sabbagh, Karl. *21st Century Jet: The Making and Marketing of the Boeing 777*. New York: Scribner, 1996.

Vincenti, Walter G. *What Engineers Know and How They Know It: Analytical Studies from Aeronautical History*. Baltimore: Johns Hopkins University Press, 1990.

CHAPTER 10

Ali, Mir M. *Art of the Skyscraper: The Genius of Fazlur Khan*. New York: Rizzoli International, 2001.

Bascomb, Neal. *Higher: A Historic Race to the Sky and the Making of a City.* New York: Doubleday, 2003.

Billington, David P. *The Tower and the Bridge.* Princeton, NJ: Princeton University Press, 1983.

Bucciarelli, Louis L. *Designing Engineers.* Cambridge, MA: MIT Press, 1994.

Buderi, Robert. *The Invention That Changed the World.* New York: Simon & Schuster, 1996.

Khan, Fazlur R. "100 Story John Hancock Center, Chicago: A Case Study of the Design Process." *Engineering Structures* 5 (1983): 12.

Lowrance, William W. *Of Acceptable Risk: Science and the Determination of Safety.* Los Altos, CA: William Kaufmann, 1976.

Lucie-Smith, Edward. *A History of Industrial Design.* New York: Van Nostrand Reinhold, 1983.

Martin, Mike W., and Roland Schinzinger. *Ethics in Engineering.* 2nd ed. New York: McGraw-Hill, 1989.

Norman, Donald A. *The Design of Everyday Things.* New York: Basic Books, 1988.

Sabbagh, Karl. *Skyscraper.* London: Macmillan, 1989.

Stoller, Ezra. *The John Hancock Center.* New York: Princeton Architectural Press, 2000.

Walton, Mary. *Car.* New York: Norton, 1997.

CHAPTER 11

Bilstein, Roger E. *Stages to Saturn.* Washington, DC: NASA, 1980.

Durant, Frederick C. III, ed. *Between Sputnik and the Shuttle: New Perspectives on American Astronautics.* San Diego: Astronautical Society, 1981.

Lewis, Richard S. *Appointment on the Moon.* New York: Viking, 1969.

Messel, H., and T. S. Butler, eds. *Pioneering in Outer Space.* Sydney: Shakespeare Head Press, 1970.

Murray, Charles, and Catherine Bly Cox. *Apollo: The Race to the Moon.* New York: Simon & Schuster, 1989.

Pellegrino, Charles R., and Joshua Stoff. *Chariots for Apollo.* New York: Avon, 1985.

Index